OBSTACLES TO RECOVERY
IN
VIETNAM AND KAMPUCHEA

U.S. EMBARGO
OF
HUMANITARIAN AID

by
Joel Charny
John Spragens, Jr.

with a Preface by
Laurence R. Simon

JOEL R. CHARNY is Southeast Asia Projects Officer for Oxfam America. In 1980, during the emergency period, he served for six months as the Administrative Officer for the OXFAM/NGO Consortium program in Phnom Penh. Since assuming his present position in early 1981, Mr. Charny has traveled to Kampuchea seven times to monitor Oxfam America's rehabilitation program.

JOHN SPRAGENS, JR. is editor of *Indochina Issues*, a publication of the Center for International Policy in Washington, D.C. He has followed events in Indochina since 1966. Between 1966-1974, Mr. Spragens lived a total of three years in Vietnam. He is fluent in Vietnamese and most recently visited Vietnam in 1980.

LAURENCE R. SIMON is Director of Policy Analysis for Oxfam America. During January 1984, he traveled to Kampuchea on a fact-finding mission to examine the impact of U.S. policies on hunger and poverty.

The views expressed in this report are the authors' and do not necessarily represent those of Oxfam America's staff or Board of Directors.

Cover Photo: John Spragens, Jr., Ha Bac Province, Vietnam. Bailing water from flooded fields. Copyright © 1980 by John Spragens, Jr.

Map designed by Jerry Alexander using facilities courtesy of Slide Graphics, Inc.

Photographs by John Spragens, Jr. Copyright © 1980 and 1983.

ISBN 0-910281-02-5

Acknowledgements

The authors would like to acknowledge the generosity of Jon Schall of South End Press in Boston who allowed us to use the page proofs of Michael Vickery's excellent new book on Kampuchea to be published in 1984. Doug Hostetter of the American Friends Service Committee gave the manuscript careful review based on his long experience related to Indochina. Efrat Levy of Oxfam America spent numerous hours in front of the word processor patiently typing the manuscript and dealing with the numerous corrections and revisions, while Oxfam America staff members Colleen Westbrook and Shari Zimble worked long hours coordinating production schedules and preparing the manuscript for publication. Mark Kelly generously donated his time and energy as a technical advisor. We give special thanks to these six, and acknowledge the help and support of our colleagues at Oxfam America and the Indochina Project.

Contents

Preface

American private voluntary organizations are being prevented by the United States government from carrying out humanitarian aid programs. Using the threat of criminal sanctions, the Reagan administration seeks to turn such organizations into handmaidens of its foreign policy.

The Trading with the Enemy Act, passed during World War I and later extended, and the Export Administration Act, passed in 1969, virtually prohibit American financial transactions, including development assistance, with four declared "enemies": Cuba, North Korea, Kampuchea (as Cambodia is now known) and Vietnam. Of these four, Oxfam America funds projects in the latter two countries. Arbitary denials of export licenses for humanitarian aid programs to Vietnam and Kampuchea suggest that the Reagan administration is pursuing a policy articulated by John Holdridge, former assistant secretary of state for East Asia and the Pacific, during his Senate confirmation hearings: to "bring the maximum political and economic pressure to bear on Vietnam... Unless, the Vietnamese feel pain, they'll have no incentive to leave Cambodia."

Most relief workers would welcome real political and positive economic incentives for the withdrawal of Vietnamese troops who swept through Kampuchea five years ago to oust the Khmer Rouge government led by Pol Pot which was responsible for the genocide of at least one million of Kampuchea's people. We would welcome a stable peace which would permit Kampuchea to commit all its resources to reconstruction and development. However, the Vietnamese are not about to withdraw while tens of thousands of armed soldiers, the majority still led by Pol Pot, wage war along the Thai border. While Oxfam America takes no position regarding a specific political settlement of the conflict, there are several overriding reasons why humanitarian organizations must not be constrained by U.S. law or its arbitary application.

As a humanitarian organization which works with the poor of about 30 countries in Asia, Africa and Latin America, Oxfam America cannot allow any government to define our enemies for us. We must be able to respond to human suffering wherever our aid is most needed. We work in countries across the political spectrum, largely through non-governmental channels, always on projects we identify and monitor. Our colleagues at the British OXFAM report they are under no legal restrictions in giving private development assistance to countries the British government has blacklisted. The U.S. government may have good reasons to restrict diplomatic relations abroad, but this does not mean that the subsistence farmers of Vietnam and Kampuchea or their toddlers with matchstick limbs are "enemies" of the American people.

Many of the same peasants in south Vietnam and in Kampuchea once received massive U.S. food aid. After years of war that severely reduced these two countries ability to feed their people, American aid—even the small private response—would heal wounds rather than inflict more pain. The gesture of caring is never lost even on those who some consider enemies.

The very choice of "enemies" has also proven arbitrary. Vietnam is an enemy, while the Soviet Union which backs Vietnam is allowed to buy millions of tons of U.S. wheat. If Kampuchea or Vietnam had the foreign exchange to be a market for American grain, their status might be defined differently.

The ill-conceived law allows for some emergency "life-saving" aid but excludes development aid.* While the Carter administration interpreted the distinction more liberally, the Reagan State Department is determined to exclude private voluntary aid except in a token amount.

"In the event of documented evidence of a large-scale natural disaster in Vietnam," a State Department policy memorandum says, the provision of a token amount of disaster relief can be considered in the light of general diplomatic situation at the time, particularly vis-a-vis Kampuchea. Generally, the statement continues, licenses will not be granted except "where foreign policy or other United States government interest are served."

The distinction between emergency needs and development aid is fallacious in a situation where massive starvation tomorrow can be prevented only by agricultural investments today. Kampuchea today is facing a 300,000 ton shortfall in rice and needs every assistance to irrigate new lands for dry season rice. Is irrigation under these circumstances not emergency assistance? Vietnam is similarly affected yet seed-processing and storage equipment (a Vietnam project aimed at expanding yields per acre and valued at $24,000) was disallowed by the U.S. State Department as not humanitarian relief even in a nation with only one-quarter acre of arable land per capita.

Such projects and others discussed in this report are clearly not aimed at challenging U.S. national security, but are rather an attempt to respond to the most basic interests of the people of Kampuchea and Vietnam.

Joel Charny and John Spragens, Jr. have documented the tragic impact of war and geo-politics on the basic needs of the people of Kampuchea and Vietnam. Among their conclusions are:

1. that the continuing conflict in Indo-China is draining development resources and further endangering the basic needs of the poor;
2. that the prospects for development in both countries are impaired by the economic embargo;
3. that the U.S. private voluntary community is severely constrained by U.S. law in the carrying out of its traditional humanitarian mission despite the support of significant numbers of U.S. citizens for its programs; and
4. that the interpretation and application of U.S. law by the Reagan administration is arbitrary and unjust and, according to legal experts, contrary to the intent of Congress.

In carrying out its foreign policy toward Vietnam and Kampuchea, the United States is losing an opportunity for reconciliation with two countries that suffered tremendous human and material losses during the war and in its aftermath. The contributions of the people of the United States through their private voluntary organizations symbolize people-to-people reconciliation. We hope that the small advances made will help open the door for diplomacy and official assistance. Our report, then, is a signpost along the way toward reversing the present trend toward the politicization of foreign economic aid.

As presently directed, food is used as a powerful weapon against governments in the disfavor of the United States. Hunger is a cruel and unworthy tool in international diplomacy. Promoting food security and development assistance are more worthy of a great nation seeking influence among the poor countries of the world.

Laurence R. Simon
Phnom Penh, Kampuchea
January 1984

*At the time of publication, the Congress was considering revisions to the Export Administration Act which would broaden aid criteria for donations intended to meet basic human needs. Though a welcome change, the PVO community would still be required to apply for export licenses through Administration channels and subject to State Department interpretation.

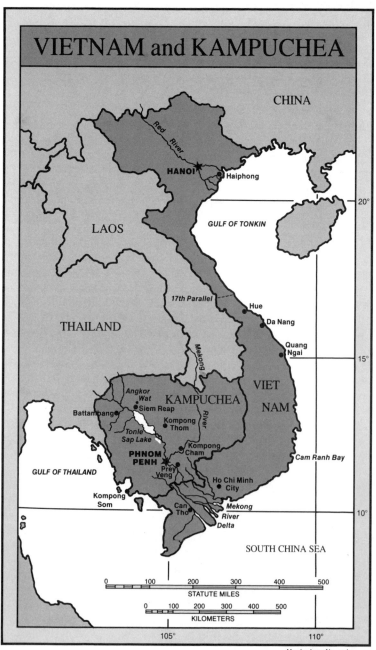

VIETNAM and KAMPUCHEA

CHINA

Red River

HANOI

● Haiphong

20°

LAOS

GULF OF TONKIN

17th Parallel

● Hue

Da Nang

THAILAND

Mekong

Quang Ngai

15°

Angkor Wat

KAMPUCHEA

VIET NAM

● Siem Reap

● Battambang

● Kompong Thom

River

Tonle Sap Lake

PHNOM PENH

Kompong Cham

Cam Ranh Bay

Prey Veng

GULF OF THAILAND

Ho Chi Minh City

Kompong Som

Can Tho

Mekong River Delta

10°

SOUTH CHINA SEA

| 0 | 100 | 200 | 300 | 400 | 500 |
STATUTE MILES

| 0 | 100 | 200 | 300 | 400 | 500 |
KILOMETERS

105°

110°

Map by Jerry Alexander

CHAPTER ONE
The Embargo

At the end of a long day of work in January 1984, the peasants of Popea Pork village in Kompong Chhnang province in Kampuchea were still harvesting. They had to bend low to gather and cut long stalks of floating rice which, when the flood waters receded, lay flat on the dry fields. There was little grain on the stalks, which made the harvest less than joyful. The crop was meagre and would not meet the grain needs of the peasant families in the coming year.

As they harvested the peasants of Popea Pork did not hesitate to communicate their problems. *Banyiha*—the Khmer word for "problem"—is one word foreign aid workers learn quickly in Kampuchea today. Production had been poor since the peasants returned to their village in 1979 when the brutal ruler, Pol Pot, was driven from the country. In January 1984, they were harvesting only half a ton of unhusked rice per hectare (one hectare = 2.47 acres), which is only half of the average yield in Kampuchea (already among the lowest in Asia) and one-fifth the yield on the same land in the 1960s. But then they had tilled their land with tractors and used locally-produced phosphate fertilizer to enrich the soil. Now, not only did the peasants have no tractors, but they had only four pairs of draft animals for 10 families. Phosphate fertilizer was unavailable. Even the rice varieties best suited to their land had been lost during the Pol Pot period. In 1983 they bought floating rice seed from neighboring Pursat province, but now at harvest time they were discovering that results with the new seed were not good.

The 1970s were devastating to the families of Popea Pork. The prosperous peasants of the pre-1970 period suffered more than physical destruction. Many were killed in the civil war, by American bombing or the fighting between government troops and the Khmer Rouge, or worked or starved to death during the Pol Pot period. Children and adolescents were taken away by the Khmer Rouge to work as part of mobile youth teams; many had never returned to their families. Now the 10 families working their land together had only 33 members. Of the 18 adults, only four were males, seven of

1

the women were widows. With poor harvests and so little labor and so few resources available to improve production, the families depended on government food aid to reach subsistence; 6 kilograms of rice and 5 kilograms of paddy (unhusked rice) per month per family were provided to the peasants of Popea Pork in 1983.

Oxfam America, along with other private humanitarian agencies working in Kampuchea, has directed its program towards meeting the needs of peasant farmers like the villagers of Popea Pork. With funds donated by thousands of American citizens, Oxfam America has supported programs designed to contribute to the achievement of food self-reliance by the peasants of Kampuchea—supply of basic farm implements, rice seeds, vegetable seeds and fertilizers; repair of a small phosphate factory and rice mills in Battambang, the country's rice bowl province; repair of large irrigation stations and small traditional irrigation pumps. These projects, along with many others supported by American agencies such as the American Friends Service Committee (AFSC), the Mennonite Central Committee (MCC), and Church World Service (CWS), have supported the heroic efforts of the people of Kampuchea to rebuild their country after the disaster of the 1970s.

U.S. government policy now threatens continuing aid to the long-suffering people of Kampuchea. For Kampuchea, under the Trading with the Enemy Act, is designated an enemy of the United States and therefore subject to a complete trade embargo. Any financial transaction with Kampuchea—along with the other "enemies" (Vietnam, Cuba, North Korea)—by an American individual, corporation or humanitarian agency must be licensed by the U.S. government. The State Department has used the power granted to the Executive by this law to deny license applications submitted by American aid agencies to ship privately donated humanitarian aid to the people of Kampuchea and Vietnam. In effect, the people of Popea Pork, and all the people of Kampuchea and Vietnam, as poor as they are and with all that they have suffered, are named enemies of the richest country on earth, the very country which has contributed so much to their agony. They become enemies not only of the U.S. government, but of all the people of the United States.

In 1983, at the very time the peasants of Popea Pork were struggling against innumerable obstacles to feed themselves, the State Department denied licenses to projects designed to help them overcome these obstacles. To meet the need for phosphate fertilizer, Oxfam America proposed to supply transport equipment to the Battambang Phosphate Factory. This would facilitate increased production by ensuring a more regular supply of locally mined phosphate rock to the crusher repaired by a previous Oxfam America grant. To allow more efficient milling of rice in Battambang, the source of surplus food rice to poorer areas like Popea Pork, Oxfam America proposed to supply materials to help repair a large rice mill powered by steam generated by burning rice husks. To help protect the health of Kampuchea's reduced draft animal population, AFSC proposed to send animal vaccines and vaccination equipment. But the U.S. State Department denied licenses to these important projects which address fundamental needs of the peasants of Kampuchea. The phosphate factory and veterinary projects were approved on appeal only after a lengthy, expensive process which delayed the provision of these key resources.

These denials are not isolated examples, but part of an explicit U.S. policy to prevent even private humanitarian assistance from reaching the people of Kampuchea and Vietnam. Since 1981 the State Department has:

- DENIED a license to MCC to ship donated Kansas wheat to Vietnam. The denial was reversed after a public outcry at this attempt to block food aid; the provision of food aid is normally held to be above political or foreign policy considerations.
- DENIED a license to MCC to ship school kits which included basic school supplies donated by American school children to children in Kampuchea. Again the denial was reversed after a public campaign featuring Mennonite school children writing letters pleading with President Reagan to allow them to donate the supplies to Kampuchean children.
- DENIED licenses to Oxfam America to ship modest amounts of seed processing and storage equipment to an agricultural cooperative and a university-based extension program in Vietnam. These denials occurred at a time when Vietnam was facing a one million-ton rice deficit.
- DENIED a license to Oxfam America to help a small bee-keeping co-op in Vietnam designed to supply honey as a food supplement to pre-school and kindergarten children. This decision was reversed only after a 15 month-process involving appeal of the denial (rejected) and submission of a new application.
- DENIED licenses to MCC and AFSC to ship emergency food and medicines to the victims of Typhoon Nancy in Vietnam. The State Department held up the applications for six months and then denied them on the grounds that the emergency no longer existed.
- DENIED a license to Oxfam America to ship 10 solar irrigation pumps to Kampuchea to help with farm-level water needs.

This list does not include projects never submitted for consideration on the grounds that to do so would merely waste precious staff time and resources. Tractor spare parts, support of a tractor repair workshop, heavy earth-moving equipment for irrigation dam and dike repair, spare parts and new equipment for further expansion of phosphate production are examples of aid which American agencies would already have sent to Kampuchea in the absence of the restrictive State Department policy.

American voluntary agencies, therefore, face not only the difficulties of contributing to solutions of the massive problems afflicting the people of Kampuchea and Vietnam, but face other obstacles to fulfilling their traditional role put in place by the U.S. government. The difficult humanitarian problems on the ground, coupled with the hostile policy of the United States, together constitute major obstacles to recovery for the people of Kampuchea and Vietnam, two countries which on every international index of poverty are among the poorest in the world.

THE LAW

The Trading with the Enemy Act, administered by the Treasury Department, and the Export Administration Act, administered by the Commerce Department, together constitute the legislation which defines the export-import controls of the U.S. government.[1] These laws, primarily designed to regulate U.S.-based commercial transactions with foreign states, are broad enough in their delegation of discretion to allow the restriction of even private donations of humanitarian assistance if deemed necessary by the Executive branch to implement U.S. foreign policy. Oxfam America's aid to Kampuchea and Vietnam has been blocked by these laws and their interpretation.

The Trading with the Enemy Act

The Trading with the Enemy Act was passed in 1916 in anticipation of hostilities between the United States and Germany. As its name suggests, the primary purpose of the law is "to give the President full power to conduct economic warfare" against "enemies" of the United States.[2] The Executive designates the enemies of this country and defines the parameters of the embargo. This perogative is especially important when the United States is in fact not engaged in a shooting war with any country; the President still has the power to impose a complete economic embargo on any nation deemed an enemy, even in the absence of direct military confrontation.

The list of current enemies (technically known as Category Z countries) is revealing: Kampuchea, Vietnam, Cuba, and North Korea. All four are socialist countries. The United States intervened militarily against a perceived communist threat in all four countries and committed thousands of troops in direct engagements with North Korea and Vietnam. With the exception of North Korea, all are poor, agrarian countries. Vietnam and Kampuchea are among the poorest countries in the world. Direct military engagement with the United States or defeat of a U.S. client state appear to be the primary criteria for designation as an enemy country. The People's Republic of China, for example, the product of the civil war opposing Mao's communists against U.S.-supported Chiang Kai-shek and the Koumintang, was a Category Z country until relations with the United States finally thawed in the early 1970s. Now the huge potential of the Chinese market makes it unlikely that U.S. commercial interests would accept any reimposition of the embargo even if U.S.-China relations deteriorate. Ironically, the Soviet Union—our presumed number 1 enemy—has never been a member of the Z club, even during the worst periods of the Cold War or after the invasion of Afghanistan. The size of the Soviet market and the lack of direct military hostilities mitigate against the imposition of a total embargo, although some restrictions on strategic goods exist, and the United States once imposed the grain embargo.

Angola, though ruled by a government firmly in the Soviet camp, is not designated an enemy of the United States. This provides further indication that previous hostilities with the United States or a client regime are the primary factor. The current Angolan leadership emerged after a bitter post-colonial civil war which did not challenge major U.S. security interests. Although the United States has provided aid to forces resisting the Marxist government, Angola has not been subject to a trade embargo. Indeed, U.S. oil companies have extensive holdings off the Angolan coast and have worked out mutually beneficial relationships with the Angolan government. Placing Angola in the Z group would end the oil companies' involvement there.

The Trading with the Enemy Act, under the jurisdiction of the Treasury Department, applies to the following types of financial transactions or commercial relations with embargoed countries:[3]

- the transfer of funds to embargoed countries;
- the purchase of goods outside the United States with U.S.-based funds for shipment to embargoed countries;
- the shipment of goods by foreign companies owned or controlled by U.S. firms;
- all financial and commercial transactions with embargoed countries, including imports into the United States;
- freezing of assets of embargoed countries in the United States.

The Act applies equally to any U.S. citizen or institution, whether established in this country or abroad. Oxfam America, IBM, the local church, and each and every citizen of the United States are subject to identical restrictions on the above transactions with embargoed countries.

The restrictions are complete with a few very minor exceptions. Relatives may remit money to their families living in Vietnam in amounts up to $300 in any three months, or $750 on a one-time basis for the purposes of emigration. Similar allowances are authorized in the case of Cuba.[4] Even this tiny crack may be sealed, however. Legislation introduced by Senator William Armstrong, Republican from Colorado, in March 1983 would prohibit *all* currency exports to Vietnam, including remittances from families in the United States to their relatives. This law would block remittances until "the President determines that the Government of Vietnam is returning to Vietnam those Vietnamese nationals that the government has sent against their will to other Communist countries," a reference to allegations that the Soviet Union is using Vietnamese "slave labor" to help build the trans-Siberian gas pipeline.[5] Senator Armstrong may hold hearings on this bill in the spring of 1984.

The Export Administration Act

The Export Administration Ace (EAA), first enacted in 1969, and up for Congressional review in 1984, legislates U.S. export controls. In contrast with the Trading with the Enemy Act, the EAA concentrates more on the terms of normal trade between countries than on economic warfare. The EAA "regulates ordinary export controls from a fluid policy responsive to political events and thus open to many exceptions," while the Trading with the Enemy Act "designates certain countries as enemies and blocks all economic intercourse with such countries."[6]

The EAA states, "It is the policy of the United States to use export controls to protect its national security, further its foreign policy, and protect the domestic economy from the excessive drain of scarce materials."[7] National security controls relate only to items which would "significantly enhance" the miltary capabilities of the recipient country "to the detriment of U.S. national security."[8] Furthering foreign policy is obviously a much broader purpose and humanitarian assistance is considered an aspect of foreign policy export controls. The restriction on exports of scarce materials is self-explanatory.

The EAA regulates the export of the following items under the jurisdiction of the Commerce Department:[9]

• goods of U.S. origin from the United States or re-exported from another country;
• components of U.S. origin to be incorporated in goods assembled or manufactured abroad;
• technical data of U.S. origin;
• goods produced abroad directly from technical data of U.S. origin.

While the language of the EAA applies to all countries, in practice restrictions are applied primarily to Eastern bloc countries and most strictly to the Category Z countries enumerated in the Trading with the Enemy Act.

The EAA leaves the power of the President to use food as a political weapon unassailed. The Act states that export controls on food "may not be imposed, expanded, or extended" if such controls "would cause measurable malnutrition." [10] But it undercuts this statement by granting the President the right to impose controls if they are "necessary to protect the national security interests of the United States" or if "the President determines that arrangements are insufficient to ensure that the food will reach those most in need."[11] The Act grants an unequivocal exemption only to medicine and medical supplies.

The Act gives the Executive the power to block development aid as well. Congress notes its concerns with this provision, but again backs off from legislating an exemption: "It is the intent of Congress that the President not impose export controls under this section on any goods or technology if he determines that the principal effect of the export of such goods or technology would be to help meet basic human needs."[12] The phrase "basic human needs," which entered the development jargon in the 1970s, suggests the right of the poor in developing countries to food, clean water, medicine, decent housing, schooling, etc. Thus, while Congress would apparently prefer to exempt small-scale development assistance from export controls, the language is modest and ultimately cedes discretion to the Executive.

Exemptions for humanitarian aid, however, "shall not apply to any export control on medicine or medical supplies which is in effect on the effective date of this Act or to any export control on food which is in effect on the date of the enactment of the Export Administration Amendment Act of 1981."[13] Thus, the complete embargoes against Vietnam, Kampuchea, Cuba and North Korea remain in place as they were imposed before the enactment of this legislation.

THE LAW'S EFFECT ON
U.S. PRIVATE VOLUNTARY ORGANIZATIONS

Despite the intent of Congress, despite the amount of U.S. government funds disbursed annually to U.S.-based international private humanitarian agencies, and despite their secure place in the American public consciousness, the work of these agencies in embargoed countries requires government approval. Oxfam America, along with several other American agencies, including AFSC, MCC and CWS, has had its humanitarian work in Vietnam and Kampuchea delayed, circumscribed, and blocked by State Department officials in Washington, D.C.

In contrast, European voluntary organizations enjoy considerable freedom from government restrictions on where and how they spend their privately donated funds. Oxfam America's sister organization, OXFAM (UK), has raised and spent millions of dollars for Kampuchea and hundreds of thousands of dollars for Vietnam, despite British government policy towards Indochina which mirrors the hardline U.S. approach. Britian provides no bilateral assistance to either country, and has lobbied the EEC (Common Market) to cut off its aid programs for Vietnam. Yet British law recognizes the distinction between government policy and the right of private organizations like OXFAM to respond to needs as they judge appropriate.

The State Department decides what the private agencies may or may not provide to embargoed countries. Although technically the responsibility for licensing lies with the Treasury and Commerce Departments, license decisions relating to embargoed countries are considered matters of foreign policy. Thus, all license applications for Vietnam and Kampuchea are referred to the State Department for "review," or what amounts to the decision itself. Two offices in the State Department review the license applications—the Vietnam-Laos-Kampuchea (VLK) Desk and the Bureau of East-West Trade, with the former assuming the primary role in the decision-making process. The VLK Desk staff, under the supervision of the assistant secretary of state for East Asia and the Pacific and his deputy, develop the embargo policy as it relates to aid. When further information is required in support of a license application (one of the more common harrassment and delaying tactics of the State Department is to make repeated requests for new information and clarification), VLK Desk staff contact the staff of the voluntary agencies directly. Thus, no one maintains the fiction that Treasury or Commerce grants the licenses; in fact, State has the final say.

8

The license applications require standard descriptions of the goods to be provided and their value. If the goods are supplied in response to a disaster, general licenses may be applied for and granted without a specific list of quantifiable items. Thus, during the Kampuchean emergency in 1979-80, agencies received multimillion dollar licenses to purchase goods which everyone, including the State Department, acknowledged were essential for Kampuchea's survival: food, medicine, rice seed, hand tools, irrigation pumps, fishnets, fertilizers, insecticides and spray sets, mosquito nets, and numerous other items. These general licenses gave the agencies the discretion necessary to respond to the emergency conditions. Outside of the urgent context of a disaster response, license applications must be more specific, more oriented to individual projects. All of Oxfam America's license applications have been submitted on a project-by-project basis, with clear descriptions of the background and rationale for the project, and summary budgets. These applications essentially duplicate Oxfam America's internal project assessment and approval procedures.

The application process is very time-consuming (see Appendix V for a complete description of Oxfam America's licensing experience). For Vietnam, the average time from application to final decision on Oxfam America's projects has been more than six months. For Kampuchea delays have averaged more than three months. Some of the delays result from the typical slowness of bureaucratic communications—the application must travel from Treasury or Commerce to State/VLK to State/East-West Trade and finally back to the original office. Given excessive workloads in each office, the possibilities of inadvertant delay are numerous. Add to this the occasional deliberate obstruction and the time needed to discuss and come to a decision and it is easy to see how the process can drag on for months. The voluntary agencies, therefore, find that their efforts to respond rapidly and efficiently to the needs of people in Vietnam and Kampuchea fall victim to bureaucratic inefficiency and obstruction.

The only State Department policy statement which has been shared with the voluntary agencies dates from February 2, 1981 (see Appendix IV for the complete text of this document). President Reagan's election coincided with the recognized end of the emergency period in Kampuchea; thus, early 1981 was an opportune time for a review of U.S. policy on licensing shipments to Kampuchea and Vietnam. The memorandum notes that U.S. government funds have been provided to Kampuchea primarily to meet emergency needs under the terms of Public Law 96-110, passed in the fall of 1979. Essentially, the U.S. Congress with President Carter's blessing agreed to suspend the trade embargo temporarily in view of the

9

desperate needs of the people of Kampuchea after Vietnamese troops and a small contingent of rebelling Kampucheans drove Pol Pot from the country. While U.S. political rhetoric tended to emphasize the alleged human rights violations of the Vietnamese invaders, in practice officials acknowledged the human needs of the Khmer people growing out of the disastrous rule of Pol Pot and the Khmer Rouge. However, the memorandum states, "as the situation in Kampuchea improves there will be a declining need for emergency shipments of food, clothing, medicine, and related supplies. There will be a corresponding need to examine more closely proposed projects involving U.S. funds to ensure that they satisfy the statutory requirement of 'providing emergency relief' and do not constitute projects for the rehabilitation or development of Kampuchea."[14]

This distinction between relief on the one hand and rehabilitation and development on the other is fundamental to the State Department's licensing criteria. In the February 2, 1981 memo the staff of the VLK Desk are unusually direct in their formulation of the distinctions as they related to Kampuchea early in 1981. Their formulation is worth quoting in full:

> Rehabilitation projects are defined here as those designed to restore the situation in Kampuchea to that which existed before the Vietnamese invasion in December 1978 (e.g. spare parts for existing machinery, boats and nets to replace those lost or destroyed, supplies to enable medical facilities previously in existence to resume operations). Development projects are defined as those designed to begin new enterprises or operate old ones at previously unattained levels. It is recognized that there are not always clear-cut distinctions between relief, rehabilitation, and development. Circumstances can alter, so that projects to supply seed, agricultural implements and nets could be justified during a severe food shortage, but would be considered to be rehabilitation projects once a level of subsistence is close, and additional food production would lead to surpluses. In circumstances where rehabilitation projects contribute toward the goals of relief, as defined above, the use of U.S. funds can be authorized on a case-by-case basis. U.S. funds cannot be authorized for development projects in Kampuchea.[15]

By this definition virtually all projects in Kampuchea since 1979 have been rehabilitation projects. Private agencies have committed resources to improve food production and rehabilitate schools, hospitals, and factories. But rice production is still only 60 percent of that attained before 1978, while schools, hospitals, and factories suffer from chronic shortages. Nothing in Kampuchea is functioning "at previously unattained levels." The goal of the voluntary agencies in Kampuchea is to assist the Khmer people in their recovery from the disasters of the 1970s. Distinctions among relief, rehabilitation, and development, which State Department staff openly acknowledge are not always clear, merely obscure the fundamental fact that the people of Kampuchea continue to require a great deal of assistance to re-build their shattered society.

The memorandum virtually rules out aid to Vietnam. "Any assistance to Vietnam, beyond the most elementary humanitarian aid, cannot be approved at this time because it would directly assist the SRV [Socialist Republic of Vietnam] in pursuing its adventure in Kampuchea."[16] Thus, government aid is impossible, except for Food for Peace commodities which could be authorized in an exceptional emergency depending on the international situation prevailing at the time of the disaster. The memo treats the issue of private voluntary assistance to Vietnam in overtly political terms:

> The private groups which provide aid to Vietnam serve the U.S. national interest in maintaining a channel of communication to the officials and people of Vietnam. The existence of this private aid program also provides a means by which the USG can, when the time arises, send a positive signal to Vietnam by permitting an increase in the level of private assistance. However, until the time comes to send that positive signal, the private aid should be maintained at a token level.[17]

At present, even private agency assistance at token levels becomes tangled in the complicated web of U.S.-Vietnam relations, relations characterized more by mutual mistrust and hatred than by reconciliation or healing of the wounds of war. When approval of small-scale humanitarian assistance projects takes on the aspect of sending a "positive signal" from one government to another, the significance of the projects themselves is blown all out of proportion. This paralyzes the bureaucracy and leads to unnecessary delays. While the people of Vietnam wait, the State Department decides what signal to send to their government. But no communication is really taking place; the only clear message is that our own government is petty and inhumane. This message the Vietnamese understand to no one's benefit.

The February 1981 memo distinguishes slightly between the use of U.S. government funds in Kampuchea and Vietnam and restrictions on privately-donated funds of the voluntary agencies. This distinction often becomes blurred in the less formal policy formulations of the VLK Desk staff. For example, in a letter to an Oxfam America donor, David Keeler of Arvada, Colorado, who had written to protest the denial of our Vietnam license applications, Desaix Anderson, the former Director of the VLK Desk, wrote, "We are unwilling to take an active role in helping Vietnam solve its internal economic and developmental problems while it ignores its obligations to its people in favor of a massive diversion of scarce resources to its occupation of a neighboring country."[18] Of course, in applying for licenses to ship small amounts of humanitarian assistance to Vietnam, Oxfam America and other voluntary agencies do not insist on an active role for the U.S. government. Yet Anderson's statement implies that even the granting of a license request involves an "active role" for the United States on the side of development in an "enemy" country. This attitude contributes further to the lack of independence for private agencies under this strict interpretation of the foreign policy function of export-import controls.

Except for an occasional letter such as the one just quoted, the State Department has put virtually nothing on paper about its licensing policy beyond the memo of February, 1981. While agencies were informed in early 1983 that there were "new guidelines" for Kampuchea, stricter than those of the original memo and closer to those in effect for Vietnam, there is nothing written to help the agencies understand the guidelines or the thinking which went into the change. This reluctance to articulate policy even for the benefit of those most affected by it underscores two key points about the licensing process: the lack of clear objective criteria for decision-making and the lack of public review of license decisions and their rationale. These factors combine to give the State Department virtually complete freedom to make judgments in keeping with the partisan foreign policy goals of the moment. For the State Department, the more closed the licensing process is, the better. Accountability is unobtainable in the atmosphere of obscurity surrounding license decisions.[19]

The few license cases which have been widely publicized indicate that the State Department is quite justified in fearing public intervention in the licensing process. The experience of voluntary agencies is that few Americans support the political manipulation of humanitarian aid, even aid to Vietnam, a country towards which most Americans are ambivalent, if not hostile. In 1981, however, as noted in the introduction, the Mennonite Central Committee (MCC) was twice able to overturn negative State Department license decisions: once in the case of a shipment of 250 metric tons of donated wheat for Vietnam and again in the case of school kits donated by young children for children in Kampuchea. The latter case was a classic and merits detailed discussion.[20]

In July, 1981 MCC filed an application with the Commerce Department seeking authorization to ship 30 tons of milk powder, 20 tons of laundry soap, and 86,000 educational kits to Kampuchea. The kits would consist of basic educational supplies—notebooks, pen, pencil, ruler, eraser—placed into small individual canvas bags for distribution to primary school children in the eastern province of

Svay Rieng where MCC was concentrating its aid to education. The beauty of the project was that MCC did not intend to purchase the supplies in bulk. Rather, school children in Mennonite Sunday schools were going to donate the school supplies themselves and put the kits together one-by-one. The project was a model "people-to-people" project.

After a four-month delay the Commerce Department notified MCC that because school supplies do not fall into the category of emergency relief, certain "negative considerations" had been raised with respect to the application. In effect, then, because the educational kits would not contribute directly to meeting the urgent food and medical needs of Kampuchea, needs which MCC was helping to meet with other projects, the license application would be denied.

This decision was based on a particularly narrow reading of what constituted emergency relief. During the Pol Pot period from 1975-79, schools, except for a minimal amount of basic literacy education, had ceased to exist. One of the most moving and significant signs of the post-1979 recovery was the reopening of schools. Children crammed into classrooms on double shifts. Most students, even those as old as twelve or thirteen, were in the first grade because of time lost under the Khmer Rouge. The Ministry of Education could not possibly provide educational supplies to the one million students who had poured back into the school system. While UNICEF had taken the lead from the beginning of the emergency in providing school supplies *as emergency assistance*, there were still severe shortages throughout Kampuchea in 1981. In this context, the project was compelling, especially since distribution of the kits would take place in one of the poorest provinces in Kampuchea, where peasants could not possibly afford to purchase these supplies for their children.

Further, when the State Department denied the license application of MCC, they were not simply denying them permission to ship school supplies. They were denying little Mennonite school children the right to aid children in Kampuchea. This human angle was eagerly seized upon by the press and articles about the denial appeared in the *Los Angeles Times*, the *Philadelphia Inquirer*, the *Christian Science Monitor*, the *New York Times*, and the *Washington Post*. Senator John Heinz, Republican from Pennsylvania, a state with a large Mennonite constituency, contacted Commerce and State on MCC's behalf. Senator Edward Kennedy wrote to the Secretary of Commerce, Malcolm Baldrige. The school children themselves wrote marvelous letters in huge scrawly handwriting directly to President Reagan. The gist of these letters was, "I have a pen and paper. The children of Cambodia do not have pen and

13

paper. Why can't I send them to these poor children, especially at Christmas time?"

In the face of this onslaught the State Department relented, packing the approved license with conditions to save face. In this case public participation in the decision-making process was crucial. Of the numerous license applications relating to humanitarian aid filed each year, few contain the exact constellation of human factors needed to mount a major publicity campaign upon rejection. While solar irrigation pumps may be just as appropriate to Kampuchea's needs of the moment, school children are not going to write to President Reagan about them. Thus, while the use of the media and public opinion can be a very powerful tool to help overturn negative license decisions, it has special and very limited application.

The staff time required to carry out an appeal is a significant drain on the resources of small agencies which must keep their administrative expenses at a minimum. To organize the school kits campaign was a full-time job for MCC's Asia program staff. To win reversal of the negative decision on their veterinary medicines project for Kampuchea, AFSC staff had to prepare what amounted to a five-page legal brief which could answer every conceivable State Department objection to a project considered essential by the agricultural technicians of virtually every agency based in Phnom Penh. Oxfam America had to submit a similar detailed justification upon rejection of the license application to ship transport equipment to the Battambang Phosphate Factory. While relieved that the licenses were finally granted on appeal, the struggle to extract these minor concessions from the State Department drains the energies of the agencies away from their primary concern: developing more effective aid programs for the people of poor countries like Vietnam and Kampuchea. The bureaucratic warfare adds one more significant hurdle to working in a region that already presents enough challenges to American agencies.

The avenue of administrative appeal is quite narrow. Under the Trading with the Enemy Act agencies can file appeals of denials, but the appeals are reviewed by the original decision-makers. The law contains no provision for independent administrative review. The EAA provides for an appellate procedure involving an independent examiner from the Commerce Department. The examiner holds a hearing which could include State Department officials in addition to the Commerce staff nominally responsible for the decision in question. The burden on the voluntary agency is to show that the denial is based on arbitrary criteria which are not part of the law itself. The chief advantage of the appeal process under the EAA is that the State Department has to go on public record in writing with the rationale for a denial. If agencies appealed enough denials, a kind of case law would be created which would provide the basis for future applications, appeals, and public debate of the licensing issue. In this way a measure of the deliberate obscurity surrounding licensing criteria would be lifted.[21] But ultimately the appeal procedure does not threaten the power of the Executive to limit the work of the voluntary agencies in Vietnam and Kampuchea.

Research by Pierre Bergeron, by the Center for Constitutional Rights, and by a law firm commissioned by Oxfam America, indicates that legal challenges to the export-import controls are doomed to failure. The constitutional right of the Executive to declare trade embargoes and enforce the controls has consistently been upheld by the courts as a part of the responsibility for foreign policy and national security. Suits challenging the application of the controls to humanitarian aid have been denied. While the lack of clear administrative procedures and safeguards in the application of the law might constitute a violation of due process, this is a tenuous basis upon which to challenge statutes relating to national security.[22]

Conclusion

The State Department's licensing policy has had a profound negative impact on the work of private humanitarian agencies in Vietnam and Kampuchea. Projects blocked under the law as "detrimental to U.S. foreign policy interests" involve aid on a small scale designed to meet the basic human needs of the Vietnamese and Kampuchean people. These needs—for food and medicine after a typhoon, for medicines to protect draft animals, for a regular water supply, for basic school supplies—are ones which must be met for all people in poor countries, even the "enemies" of the United States. No government, and especially not the American government, should have the right to prevent private humanitarian agencies, expressing the concern of thousands of private citizens, from meeting these needs.

The value of Oxfam America's projects denied in 1982 and 1983 by the State Department is $199,062. This figure is but a tiny fraction of the U.S. expenditure on the war effort in Vietnam. Therefore, while restrictions on private assistance threaten the programs of the voluntary agencies in Vietnam and Kampuchea, they are but one component of the massive impact that U.S. foreign policy has had on these countries. To block a $13,000 seed coop project is one thing. To spend at least $140 billion to wage war in a poor country and then refuse to provide promised postwar reconstruction assistance is quite another. In the next two chapters we will examine economic development and the humanitarian situation after the Second Indochina War in Vietnam and Kampuchea in the context of the U.S economic blockade against these countries.

---

1. The basic reference for this chapter is a study commissioned by Church World Service: Pierre Bergeron, *Export-Import Controls: Their Nature and Accountability, Approaches to Fostering U.S. Recognition of Foreign Governments* (New York: Church World Service, 1980).

2. Bergeron, *Export-Import Controls*, p. 38.

3. Ibid., p. 14.

4. Ibid., p. 40.

5. Senate Bill 747, introduced by Senator Armstrong on March 10, 1983.

6. Bergeron, *Export-Import Controls*, p. 33.

7. Ibid., p. 9.

8. Ibid., p. 19.

9. Ibid., p. 12.

10. Export Administration Regulations, Legislative Authority (Section 6), Subsection (f).

11. Ibid., p. 12.

12. Ibid.

13. Ibid.

14. See Appendix, pp. 136-137.

15. Appendix, p. 137.

16. Appendix, p. 138.

17. Appendix, p. 138.

18. Desaix Anderson, Letter to David Keeler, December 11, 1982.

19. See Bergeron, *Export-Import Controls*, p. 53.

20. The following account is based on correspondence supplied by MCC staff and discussions with their staff during their fight to overturn the negative decision.

21. Bergeron, *Export-Import Controls*, pp. 63-66.

22. Ibid., pp. 66-83.

CHAPTER TWO

Vietnam

In 1984, nine years after the end of its French and American wars, Vietnam faces a complex set of difficulties. While it is no longer possible to blame its problems entirely on the destruction and social dislocation of 30 years of war, it is hard to find any problem which is not at least partly rooted in the war or complicated by its after-effects or present-day U.S. attempts to isolate and "squeeze" the Vietnamese economy.

The United States could have altered Vietnam's postwar prospects by living up to its pledge to help heal the wounds of war, but it is difficult to project a picture of what Vietnam would be like today if that one factor were changed. Some observers, such as the *Far Eastern Economic Review*'s Nayan Chanda, say Vietnam has been slow in absorbing any aid except cash grants and commodities. The lack of trained technicians, engineers and managers makes it difficult for the country to take on large industrial projects, although there are more hopeful examples at the low end of the technology spectrum. UNICEF has reported good success with its program of helping design and build model-day care centers and Oxfam America has applied for licenses to fund several small-scale projects in Vietnam, such as the Tich Giang Seed Cooperative.

The seed cooperative, which lies a few kilometers outside the Hanoi city limits, is responsible for producing high quality rice and vegetable seeds and supplying them to peasant farmers in the area. By assigning this specialized task to the Tich Giang Seed Cooperative, the Hanoi administration, in cooperation with the Ministry of Agriculture, expects to improve seed quality and thereby increase food production in the green zones around Hanoi.

The cooperative, established in 1976, cultivates 40 hectares of land divided evenly between rice seed production and production of maize and soybean seeds. The cooperative has 200 members, 90 percent of whom, including the Director, are women. The members derive income in kind by exchanging seed for paddy (unhusked rice) with neighboring cooperatives, receiving 10 percent more paddy than seed in the exchange. In addition, they sell the seed they produce for 1.5 *dong* per kilogram (less than 15 cents per kilogram at the official exchange rate). The workers also receive a low salary of 20 *dong* (about $2.00) per month. At best they are able to meet only their subsistence needs.

In January 1982, Oxfam America staff visited this cooperative as part of an assessment of the possibility of funding small-scale projects in Vietnam. The cooperative had virtually none of the equipment necessary to do an efficient job of seed production and selection. The members were drying the seed outside on concrete, a difficult task in the rainy winter months in northern Vietnam. They were storing the seed in a small warehouse where losses due to dampness or heat were common. They had no way to clean or purify the seed. The members depended entirely on manual means of production which demanded a great deal of labor and which involved significant losses.

The cooperative's modest request for assistance included seed driers, a small cold storage unit, a germinator, a microscope, and scales for weighing the seed. This would help meet the cooperative's goals of improving seed quality and thereby raising the yield of peasants using their seed. Peasants would enjoy yield increases of up to 30 percent using the high quality seeds produced by the cooperative rather than seeds gathered from their own fields.

With Vietnam facing a food deficit of nearly two million tons in early 1982, the focus of Oxfam America's program was logically to aid local efforts to increase food production. The Tich Giang Seed Cooperative seemed to be a worthwhile first-time initiative given the project's focus on increasing food production, the modest size of the request, and the proximity of the cooperative to Hanoi, which would allow for easy monitoring. Oxfam America's Projects Committee agreed, and on April 23, 1982 approved a grant of $13,000 to the Tich Giang Seed Cooperative. At the same time they approved four other projects supporting food production and community-based health programs located in Hai Hung province southeast of Hanoi, and at Cantho, in the Mekong Delta.

The U.S. government—specifically the State Department—did not allow Oxfam America to implement the Tich Giang project. Nor has our government permitted Oxfam America to meet its other commitments to support food production projects in Vietnam.

United States policy since the end of the war in 1975, and especially since Vietnamese troops went into Kampuchea in 1978 to overthrow the Pol Pot regime, has been to squeeze Vietnam's economy in an attempt to force it to change course. However, rather than slowing down Vietnam's recovery and growth, the United States may be simply reducing the range of countries and private development agencies providing aid to Vietnam—further limiting Vietnam's choice of development models.

THE AFTERMATHS OF WAR

Social and Economic Background

The social and economic impact of a war spanning thirty years is difficult to imagine. It is even difficult for someone who saw it firsthand to remember ten years later. The United States has never lived through a comparable period. But for Vietnam, the war against the French began as soon as World War II ended in 1945. The fighting continued, with occasional intervals of uncertain peace, until the war which pitted the North and Provisional Revolutionary Government (PRG) against the United States and Saigon government ended in 1975.

South Vietnam: Saigon Government Zone

Refugee camps became a fixture of provincial capitals and district towns throughout southern Vietnam. Some of the refugees had, on their own, fled combat zones. Others were deliberately removed from their homes as part of a policy of drying up the human "ocean" in which the guerrilla "fish" of the National Liberation Front (Saigon and Washington called them the Viet Cong) lived.

Those who moved on their own often went uncounted. The official refugees, who applied for assistance underwritten by the United States, numbered in the millions. American statistics on the period from 1965 to 1973 indicate more than 10 million Vietnamese were made homeless at one time or another during the war.[1] The majority of them passed through the refugee camps—rows of crowded tents or thatched sheds surrounded by barbed wire, where whole families might be assigned spaces as small as 10 square feet. These camps were symbols of despair, of the uprooting of an entire society. In a land of 19 million, which in more normal times was 85 percent rural, 40 percent of the population lived in urban areas by 1973 (even though, by then, the population of refugee camps had been reduced to 300,000).[2]

Saigon drew an especially large number of people. Even in countries at peace, cities are glittering magnets. The attraction was amplified in wartime Saigon by the free flow of American dollars (more than $400 million a year spent locally in South Vietnam between 1966 and 1971) and the fact that Saigon was relatively immune from fighting.[3] So a city designed in French colonial times for perhaps half a million was by the end of the war bloated with a population of four million or more.[4] Once refugees left the government-run camps, they were considered resettled, even if they had only joined the new urban poor who were a large and conspicuous part of the population of Saigon and other major population centers. They found shelter in flimsy shacks on stilts built out over rivers which had turned into open sewers or in the huts of gaudy metal sheets printed for soft drink cans (but never cut and shaped) which jammed the streets and alleys of working quarters of the city. Their counterparts in the countryside could be seen huddled along the shoulders of the highways in homes fashioned from whatever they could find — even the cardboard cartons discarded by American troops in the field.

Many of the refugees eked out a living on the fringes of the economy's service sector. One-time farmers became petty traders, part of a free-wheeling black and gray market that could turn up anything from a jar of Smucker's cherry preserves to an M-16 rifle or American mortar.[5] For the potential investors of the South, industrial investment held little attraction. They could make more money faster dealing on the black market, or investing in the legitimate service sector—from snack bars to real estate—which accounted for more than 50 percent of the South's gross national product as late as 1973.[6]

Not surprisingly, the industrial sector was stunted, accounting for only 10 percent of GNP in 1973.[7] The textile industry was the largest. Cement, paper, glassware and pharmaceuticals were also important. But these industries were heavily dependent on imported raw materials. There were essentially no domestic sources of fiber for the spinning and weaving mills. Patrick Boarman, of Pepperdine University's Center for International Business, noted a similar situation in the paper industry: "South Vietnam . . . has 14 million acres of prime forest, but domestic paper mills supply only one-half of the country's demand for paper and nearly all wood pulp for these mills has to be imported. The degree of reliance on imports is perverse, inasmuch as the country has all the raw materials and the labor necessary to produce pulp and paper. What is lacking is investment capital."[8]

In a similar manner, the country—especially Saigon and the other cities—had been accustomed to large-scale imports of luxury consumer goods, paid for with the flood of American dollars rather than with the country's own earnings. The surface affluence was part of the war effort, seen as a demonstration of Saigon's superiorty over its opponents. "We flooded this place with imports and forcibly raised demand and consumption way beyond what the Vietnamese could ever afford to pay," one U.S. official told *Newsweek* in 1973. "We are trying to buy votes, not solve economic problems. Now, the chickens are coming home to roost."[9] Imports in 1972 cost more than $740 million, while exports earned less than $24 million.[10] The World Bank estimated South Vietnam would need aid over a long period, with more than $11 billion in economic assistance between 1974 and 1990 simply to keep it a "viable state with a modest reconstruction program."[11]

Saigon had hoped for increased western aid and investment after the signing of the peace treaty in January 1973. In 1972, the Thieu government announced a generous investment law. Subsequent attempts to drum up investments always included references to inexpensive labor. Wages were said to be about one-third those in Singapore, four-fifths those in South Korea.[12]

Potential investors and potential aid donors were cautious. They were skeptical about prospects for a real peace, and their skepticism seemed to be supported by the U.S. Overseas Private Investment Corporation, which declined to underwrite investments in South Vietnam.[13] American companies hesitated to invest where OPIC feared to tread. The best-known factories opened under the investment law were both Japanese—one assembling Panasonic radios and the other making Kubota tillers.

The one area where American companies were conspicuously involved was in exploring for oil on the continental shelf off Vietnam's southern coast. Saigon was offering terms which Boarman characterized as a "bargain package" by comparison with those available in other oil producing countries.[14] Companies would have to pay 12.5 percent royalty on oil produced and sold and 55 percent income tax on company profits. American-owned Mobil and Exxon were among the four successful bidders for the first round of exploration, along with Shell (American, British and Dutch) and Sunningdale (Canadian). By late 1974, there were indications of oil and gas in test wells, but no producing wells had been drilled by the end of the war.[15]

While Saigon's politicians and businessmen were hoping for oil, poor people in the South faced acute hunger during the last year of the war. The causes of the hunger included a tangled combination of war, a weak economy, deliberate exploitation and corruption. Although there was a government program to encourage refugees to return to farming, in practice the program was severely limited because of political considerations. Thieu insisted that his was the "sole legitimate government" in South Vietnam, even though the 1973 Paris peace agreement had given equal status to his "Viet Cong" adversaries, the Provisional Revolutionary Government. So, although the peace treaty allowed civilians freedom of movement, in practice Saigon allowed farmers to return only to areas under the control of their troops, the Army of the Republic of Vietnam (ARVN). This severely limited the scope of the resettlement.

Where farming was going on, many farmers under the encouragement of U.S. Agency for International Development had introduced "green revolution" rice strains which required plenty of fertilizer and insecticides. One of the most publicized scandals of 1974 involved systematic hoarding of the vital fertilizers. A bag of fertilizer was selling for as much as a bag of rice. There were warnings that the hoarding could cause a shortfall of 50 percent in the next rice crop.[16] Rice (and wheat flour) were still being imported with U.S. aid (in 1973 there had been more than 300,000 tons of rice imports plus more than 200,000 tons of wheat flour).[17] This did not mean people could afford the "official" price, and opposition members of the National Assembly collected numerous stories of people living on the edge of starvation.

Phan Xuan Huy, a deputy from the major central Vietnamese port of Danang, told of meeting a 60-year old man whose son had been killed in the war. His daughter-in-law was trying to support him as well as her three children.

"When he met us, Mr. Loi had been hungry for some time," Deputy Huy reported. "Saliva ran from the corners of his mouth and he couldn't speak. When we asked him questions, his arms and legs trembled. The person next door told us that Mr. Loi had been eating only one meal of thin rice gruel a day, and sometimes only once every other day. When his daughter-in-law could sell some firewood or catch crabs, they might be able to have a meal of rice."[18]

In that one hamlet, Deputy Huy found 25 people in similar circumstances. In Danang and elsewhere, there were reports of deaths. Some were caused by accidental consumption of poisonous plants, foraged in the wild so the hungry would have something to fill their stomachs. Other deaths were deliberate—desperate parents poisoning themselves and their children because they could no longer afford food for the family.

South Vietnam: PRG Zones

Meanwhile, there was another South Vietnam, seldom seen by American reporters—the South Vietnam of the Provisional Revolutionary Government (PRG). It is more difficult to assess the economy of the PRG zone. If Saigon's economic statistics had to be treated with caution, then the PRG simply did not publish statistics for its whole zone of control. There were some occasional numbers on individual villages or provinces, but they were most often expressed in percentages rather than specific quantities or values.

A survey of various issues of *South Vietnam in Struggle*, the PRG's official foreign-language publication, indicates that government services were concentrated in the fields of education and public health. There were also a number of newspapers and publications, and itinerant performing troupes. And in at least some areas, there was a rudimentary agricultural extension service. Some of the expenses of these government services and of the army were covered through taxes or "donations" from farmers in the zone. But it seems likely that most of the military expenses and at least some of the civilian expenses were covered by North Vietnam.

If there were no systematic statistics on the economy of the PRG zone, there were some impressionistic sketches provided by visits of western correspondents in 1973 and the early part of 1974.

In the district of Que Son, a few miles south of the central Vietnamese port of Danang, the major impression was of a spare but adequate economy based as much as possible on principles of self-sufficiency. As was true in most of the mountainous central part of the country, rice was supplemented with sweet potatoes and manioc. Rice fields were in the valleys. Tubers, vegetables, peanuts and tobacco would grow on smaller, drier fields carved into the mountains. Edible gourds and melons grew on arbors set up beside thatch farm houses. Most families seemed to have at least a few chickens, but meat was an infrequent part of the diet.[19]

Farm families held and worked their land privately. They shared snacks and tobacco with local PRG officials, but were paid rent when their houses were used by officials (as shelter for visiting reporters, for example) and were paid for any food the officials used. Some government services, like schools, were provided on a regular basis. Others were *ad hoc*. Soldiers with free time, for example, had built a major irrigation aqueduct to divert stream water to upland fields.

There were markets and small shops where farmers could convert their surplus into consumer goods "imported" from the Republic of Vietnam (Saigon) (RVN) zone. The only major items in evidence were transitor radios—from Japan. Shops had batteries, tinned milk and meat, notebooks and pens and other such inexpensive items. The markets and shops were private enterprises.

An official, noting the relative lack of dependence on imports, pointed out that no petroleum products were used except for tiny kerosene lamps which provided the only light at night. Although some high-yield rice strains had been introduced, farmers used biological fertilizers for them.

Farther north, in Quang Tri province just below the north-south demarcation line, society was more organized. The proximity to the North helped, and the PRG was eager to leave a positive impression with the foreign visitors who were frequently brought there as an extension of a trip to North Vietnam.

One feature of the economy of the PRG-controlled area of Quang Tri was a state market. Some basic items sold at prices below the free market. Quantities were rationed—18 kilograms of rice per month for an adult, for example—and some items, such as cotton cloth, were available only in the state market. Resettlement assistance was provided for a full year for returning farmers, with 15 kilos of free rice a month for adults. Property was still privately held, but mutual aid groups were being encouraged. Crops could be sold to the government or on the free market.[20]

With sporadic fighting continuing in many parts of the PRG zone, there were limits to the amount of deserted land which could be reopened for cultivation. In any case, it was a difficult and dangerous task, especially in the central provinces where fighting had been most intense. Officials in Quang Tri said it took a hundred work days to remove the mines and other unexploded ordnance from a hectare of rice land there.[21]

North Vietnam

In the North, the Democratic Republic of Vietnam (DRV) was able to begin postwar reconstruction by early 1973. For two more years, the economy was burdened with the expenses of supporting the war in the South—expenses which were never publicly detailed. But at least the territory of the North was no longer under attack.

Northern Vietnam had a better industrial base than the South because the French, in the colonial period, had concentrated more of their limited industrial development there. At least in part, this can be attributed to the ready source of energy provided by the coal mines at Hon Gai, near the northern port of Haiphong. Electric power plants were fired with this coal. Other plants produced cement, fertilizer, cloth and paper.[22] After the end of the French war in 1954, these plants were restored and expanded, and some new industries such as steel making and machine shops were developed.

Like other socialist countries, the DRV adopted the five year plan as a basic tool of economic management. The first five-year plan ran from 1961 through 1965, but then the American war intervened and the long-term plans were temporarily abandoned. Some factories with relatively portable machinery were relocated from towns to protected areas in the countryside and continued to operate. Large fixed installations like steel mills or major power plants were especially vulnerable to the wartime bombing. Vietnamese economists say production in some areas, such as coal, cement, sugar, and *nuoc mam*, the fish sauce which is the basic Vietnamese condiment, fell by as much as 30 to 50 percent.[23] But they say that, thanks to a stress on regional industries, production levels in many areas were sustained. For example, between 1965 and 1970, production of cloth declined about 11 percent, from 100.3

million meters to 89.4 million. Production of farm implements rose by more than 50 percent.

Agricultural production between 1966 and 1971 also rose, but only marginally—a total of about three percent over the five year period.[24] Considering that population rose at an average rate of 2.9 percent per year between 1960 and 1974, however, the country's ability to feed itself slumped during the war years.[25] It is generally believed that China supplied the North with up to half a million tons of rice a year during the war—roughly the same as that supplied to the South by the United States—in order to cover these growing deficits.

After the 1973 peace treaty went into effect, the DRV began an interim economic plan which would run through 1975. This meant that in 1976 the country could begin a new five year plan at the same time as the Soviet bloc, which would make it easier to coordinate plans for aid, trade and other forms of economic cooperation.

Reporting on the first 19 months of this interim plan, Prime Minister Pham Van Dong said, "It is probable that by the end of 1974, the gross national product in North Vietnam will have reached or even surpassed the pre-war level."[26] Major tasks had included bringing factories back to permanent sites, beginning to fill the millions of bomb craters which pocked the countryside, and especially restoring the transportation network which had been a primary target of U.S. bombing. Getting back to the 1965 gross national product represented a considerable effort. But since the population had grown more than 25 percent during that period, per capita output would have been only about 80 percent of the pre-war levels.

So in 1975, when the war finally ended throughout Vietnam, North and South were both in a precarious economic condition.

Ecological Background of Agent Orange

"Ranch Hand," the operations were called in the picturesque jargon of the war. General William Westmoreland, who was Vietnam field commander from 1964 to 1968, referred to the program in a list of "some of the most imaginative and successful expedients and innovations to cope with the unusual nature of the enemy and the war that any military force has ever brought to bear."[27]

"Chemical defoliants . . . helped to deny the enemy hiding places, particularly ambush sites along roads and waterways. Early in the war defoliants were also used to deny the enemy rice in remote areas that were VC-controlled," he wrote in his memoir of the war. "Some ecological damage may have resulted from the defoliants; how much and how permanent it is remains to be seen. Flying over much of the country as recently as 1972, I found Vietnam still a verdant land, which left me to question the truth of some of the more pessimistic allegations of permanent damage. The defoliants were a major factor in reducing the number of ambushes that were long so costly in American and South Vietnamese lives."[28]

Medical Effects

In retrospect, Westmoreland's optimism seems startlingly glib. In a class action suit filed in federal court in New York, 20,000 American veterans of the Vietnam war are seeking damages from the companies which manufactured the defoliants. They blame the chemicals for medical problems including cancer, genetic damage and the birth of deformed children. Similar problems have been reported in Vietnam. And in Vietnam, the ecological effects have proved to be severe and long-lasting. They have affected farming, fishing, public health, and even the climate in many regions of the country.

The medical complaints reported by both American veterans and Vietnamese are attributed to dioxin, a potent poison formed during the manufacture of 2, 4, 5-Trichlorophenoxycetic acid, or 2, 4, 5-T. Agent Orange, used to defoliate trees in Vietnam, was a combination of 2, 4, 5-T and another defoliant, 2, 4-D. Between 1961 and 1969, 44 million pounds of 2, 4, 5-T were used in Vietnam.[29]

Current and potential legal actions have colored the information published about the effects of dioxin. In the veterans' suit, for example, the veterans' lawyers are seeking to prove that the manufacturers knew of the dangers of the chemical but did not inform the government.[30] The chemical companies must try to prove the government did have knowledge of the dangers—at least as much as was available. For obvious reasons, that information was based on animal studies. It is difficult to know how closely reactions in various animals correspond to reactions in humans to an untested substance. But by early 1965, at least one study had shown liver damage in rabbits.[31] A second-hand account of that study said liver damage was severe and that "a no-effect level based on liver response has not yet been established. Even vigorous washing of the skin 15 minutes after application will not prevent damage and may possibly enhance the absorption of the material."[32] That same year, tests of 2, 4, 5-T on mice showed that small doses cause birth defects.[33] (Defense Department officials learned about the report in 1969.)

At least as early as 1969, Vietnamese journalists linked defoliants to birth defects. *Tin Sang*, an opposition daily in Saigon, reported what it called "egg-bundle-like fetuses."[34] The following year, a commission of the American Association for the Advancement of Science surveyed records of abnormal births at Children's

26

Hospital in Saigon and found a sharp rise in the incidence of cleft palates and spina bifida after heavy defoliant spraying began in 1966.[35] And the commission found that stillbirths in heavily-sprayed Tay Ninh were 64 per thousand—more than twice the average for South Vietnam as a whole.

Since the end of the war, a small group of Vietnamese doctors has continued to monitor evidence of the effects of Agent Orange. These physicians, until recently led by the late cancer specialist Dr. Ton That Tung, acknowledge that they are working from limited data. In a country where it is a struggle to provide basic public health services, few people are available to work on such research projects, and they must carry on their research as one among their many responsibilities. They also lack the computers which, in a country like the United States, could help them compile and correlate data from broader surveys of people exposed to defoliants during the war.

Surveys in North Vietnam found that men whose military service had taken them to areas of the South were more likely to father children with birth defects than were men who had remained in the North.[36]

In the town of Yen Bai, for example, the researchers looked at 3,058 births between 1975 and 1978. Out of a total population of 32,000 there were 700 veterans who had served in the South. No figures are given for ages of the veterans or age and sex of the total population. But if Yen Bai is similar to the total Vietnamese population, about 4,500 women would be expected to fall in the prime childbearing ages of 20 to 40.[37] (Women in this age group account for more than 80 percent of births in Vietnam.) Thus even if all the veterans with service in the South were married to women in this age group, they would account for only about 20 percent of the total. The survey found 30 children born with deformities. Half of them were fathered by the veterans of southern combat.

Six of those children were born with no brain. All were from the veterans' group. Other defects have included the absence of nose, eyes or ears, and deformed limbs. Other Vietnamese surveys have found abnormally high rates of spontaneous abortions in areas exposed to defoliants.[38] Studies of chromosomes in mothers subjected to defoliant sprayings and in their children, both mothers and children showed unusually high rates of abnormalities, suggesting that the effects of the chemicals may be passed on from one generation to another.[39] Similar problems with birth defects and miscarriages have been reported by exposed war veterans in Australia as well as the United States.[40]

27

Ecological Effects

If the medical effects of defoliants are still poorly explored and if those effects have shown up so far in only a relatively small number of people, the environmental effects are starkly obvious and their impact is felt by the entire population of regions subjected to heavy defoliation. In some areas the devastation would be obvious even from the air. In other areas, while the countryside might have seemed "verdant" as General Westmoreland flew over it, the original plants never recovered. Instead, new varieties of plants grew, and they have very different effects on the environment—effects which may extend far beyond the immediate defoliated areas. Many of the wartime predictions of environmentalist critics have been borne out by events, as can be seen by comparing a study compiled by John Lewallen just after the defoliation program was ended with a review of developments in Vietnam recently published by Dr. Nguyen Khac Vien, a French-trained physician who has for many years directed the Foreign Languages Publishing House in Hanoi.[41]

Some of the most desolate stretches of Vietnam today are the defoliated mangrove forests. Mangroves are actually several different varieties of trees which thrive in areas like the Mekong Delta where new land is gradually being formed as silt settles out of the river. The mangroves rise on stilt-like roots which help capture more silt and other floating debris to consolidate the new land. The initial establishment of a mangrove forest takes some 20 years. The highest land in these areas is barely above high-tide level. People live in inland areas of the region in houses on stilts built along the rivers and canals which crisscross the Delta. Living mangrove forests provide rich harvests of fish, shrimp, birds, honey and wood often used to make charcoal. "With only rudimentary fishing gear," Vien reports, "local fisherman take a 200-kilogram catch each night on an average."

By the end of 1970, about half of the South's coastal mangroves had been sprayed with defoliants. American surveys indicated one application of herbicides killed 90 percent of the affected mangroves. Vien reports that more than 380 square miles, or nearly one-third of the southern mangrove forests, were affected by defoliants. About 45 percent of the affected area showed serious damage. Decaying plant matter robbed the water of oxygen, sharply reducing the number of fish, shrimp, and crabs. In some places, catches fell to 10 to 15 percent of what they had been before the defoliation. There has been some new forest growth, but about 30 percent of the area still lacks vegetation. The most serious effects are on the relatively high ground—that is, above the water level—which was exposed to sun and rain. Soil in these areas has hardened and become more saline. For two years there were efforts to grow soy beans in these areas, but the soil dried out and became more saline, and these efforts had to be abandoned. Now the emphasis is on artificial restoration of the mangroves, mainly a variety known as *Rhizophora*, which was one of the dominant types in mature, naturally-occuring mangrove forests. Vien says that by 1981 *Rhizophora* had been reestablished on about 20 percent of the defoliated area. Another 10 percent of the area—that lying close enough to surviving forests that seeds could be propagated naturally—has shown natural regeneration, and trees are six to 10 meters tall. This compares to height of 20 to 30 meters in a mature forest.

Vien gives no estimate of how much of the forest can be recreated or how long the process may take. Optimistic American predictions during the war years suggested the forests could be restored in 20 years, but these predictions were based on the assumption that seeds would immediately be redistributed to defoliated areas and take root. In fact, it was five years and longer before reseeding began.

The impact of defoliation on the dense tropical forests of the central Vietnamese mountains was equally striking. The composition of these forests had remained stable for thousands of years. In the evergreen tropical forests, often referred to as triple-canopy jungles, the tallest trees reached heights of 50 meters.[42] The lowest level of this type of forest is shrubs of two meters or less. Little light penetrates to the forest floor. In the foothills of the central mountain chain are slightly less dense forests, where the lower levels include bamboos. Because the tall trees limit the light reaching the lower levels, growth of the bamboos is limited.

Central Vietnam depends for much of its food on a narrow coastal plain where carefully tended rice paddies with good irrigation can give some of the highest yields in the country. The upland forests are important to the climate of the coastal farmlands. In the summer, winds from India drop their rains in Laos, on the western side of the mountain chain. Intact forests moderate the strength, temperature, and dryness of these winds.[43] But in the Quang Tri area, just south of the former demarcation line between North and South, many of the forests were completely destroyed by a combination of repeated defoliation, bombing and resulting fires. American and Vietnamese photographs of the Ho Chi Minh Trail, the military supply route from the North to the South, show vast areas where nothing is standing but the denuded trunks of the largest trees. As a result, summer temperatures in the coastal areas, which once were in the upper 80s and lower 90s, now reach well into the 100s. Then during the rainy season, flooding is worse because the forests are not there to hold the water.

The new growth of brushy plants and grasses, especially the

small bamboos and impereata grass which are characteristic of savannah in Vietnam, chokes out seedlings of the trees which used to dominate these forests. Vien reports a 1981 survey of the A Luoi region in this Quang Tri area. "No sizeable mammal of any value was found," he said, "while in the neighboring regions, which had not been sprayed, about fifty species were counted" including deer, boars, flying squirrels and wild cats. Only small rodents remained and prospered, bringing with them an increased risk of plague. "The regeneration of the flora and fauna has proved to be well-nigh impossible," Vien concludes. "At any rate it will call for immense efforts and a very long time."

UNITED STATES PLEDGE OF ASSISTANCE

The agreement which was to have ended the United States' war in Vietnam was signed in Paris in January 1973. United States troops were withdrawn shortly afterward, but the war dragged on for two more years, and a number of the provisions of the Paris agreement were never carried out. Among those unfulfilled provisions was Article 21, which includes this pledge: "In pursuance of its traditional policy, the United States will contribute to healing the wounds of war and to postwar reconstruction of the Democratic Republic of Viet Nam and throughout Indochina."[44]

There was a great deal of secrecy surrounding the negotiations that led to the Paris peace agreement. The "official" peace talks involved delegations from the United States, the Democratic Republic of Vietnam (Hanoi), the Republic of Vietnam (Saigon), and the Provisional Revolutionary Government of the Republic of South Vietnam (the "Viet Cong"). But the serious work which eventually led to the peace agreement was done in secret negotiations between Henry Kissinger, President Richard Nixon's national security advisor, and Le Duc Tho, a senior member of the Political Bureau of the Lao Dong (Workers) Party, the name used at that time by the Communist Party in North Vietnam.

Secret Agreements

Besides the published peace agreement, those secret negotiations also led to at least one secret agreement, which was finally disclosed by the Vietnamese three years later. That agreement was contained in a letter from President Nixon to Prime Minister Pham Van Dong of the DRV, dated February 1, 1973. That letter promised postwar reconstruction assistance from the United States to the DRV "in the range of 3.25 billion dollars."[45] Details of the aid were to be worked out by a joint commission with an equal number of representatives from the two sides. Nixon wrote that the U.S. aid would be based on such factors as "A - The needs of North Vietnam arising from the dislocation of war —and B - The requirements for post-war reconstruction in the agricultural and industrial sectors of North Vietnam's economy." An attached note on other forms of aid said, "United States studies indicate that the appropriate programs could fall in the range of 1 to 1.5 billion dollars depending on food and other commodity needs of the Democratic Republic of Vietnam."

The joint commission did discuss details of this aid, in meetings in Paris which began March 15, 1973.[46] It continued its discussions until July 23. A detailed aid agreement was to be signed at that time. Instead, the United States broke off the talks.

This hidden chapter of the peace process remained secret until Phan Hien, deputy foreign minister of the DRV, revealed it to a delegation of the House Select Committee on Missing Persons in Southeast Asia in December 1975. Rep. Paul N. McCloskey, Jr. told American newsmen of the delegation's findings, and the State Department later confirmed the existence of the Nixon letter.[47] The texts of Nixon's letter and Dong's reply were made public by the Vietnamese in May 1977.

Among the documents given to the House committee delegation in 1975 was a detailed list, eight single-spaced pages long, specifying the forms U.S. aid would take. Most of the aid money—85 percent— would be spent in the United States. The remaining 15 percent would be used to buy goods from third countries. A staff memorandum to the House Select Committee outlined the aid package described in that list:

> . . . the United States was expected to play a central role in the reconstruction of North Vietnam, with the emphasis on industrial plants and commodities, infrastructure, and energy. The five-year plan provides for plants for prefabricated housing, plumbing fixtures, sanitary porcelain ware, cement, sheet glass, chipboard, synthetic paint, and a steel mill with an annual output of one million tons. The contribution to energy development included a thermal power station with a capacity of 1,200 megawatts, a high tension electrical equipment plant with an annual output of 3,000 tons, and 20,000 metric tons of high tension copper cable. In addition, the agreement included a provision of a vast array of equipment for port reconstruction and water, road, and rail transport, and for agriculture.[48]

Any assessment of the impact of the absence of that aid, which would have totaled more than $4 billion, would be mere speculation. The types of aid outlined were eminently practical, and could have made a significant contribution to meeting some of the basic needs

31

of postwar Vietnam. The degree of the impact would have depended on the country's ability to absorb the aid. Both during and since the war, reporters have noted Soviet bloc complaints about inefficiencies in the Vietnamese use of sophisticated technical aid and about the difficulty of moving materials through Vietnamese ports. Even allowing for inefficiencies, the projected U.S. assistance would certainly have made Vietnam's recovery from the war easier in a material sense. Perhaps more important, Vietnam would have benefitted from a more favorable international political climate if the United States had been providing economic assistance rather than continuing the war by economic means.

To give some measure of the significance of the $4 billion-plus in proposed aid, it may be useful to compare it to the more than $11 billion in non-military aid which the World Bank had recommended for South Vietnam over a similar period of time,[49] or to the $6 billion in material losses the North Vietnamese said they had sustained up until July 1972.[50] This figure did not include the losses from the large-scale bombing raids of December 1972. Estimates of the total amount the United States spent to fight the war in Vietnam range from $140 billion to $676 billion.[51]

U.S. Embargo of Aid

By April 1975, when Saigon's army collapsed so quickly that the advancing forces of the North and the PRG could not keep up, Nixon had resigned and Gerald Ford was President of the United States. Henry Kissinger remained secretary of state—a post he had held since September 1973. Kissinger's State Department moved quickly to impose what amounted to an economic embargo on all of Vietnam, extending sanctions already in effect against "the Communist controlled areas" of Vietnam to the whole of the country.[52] On April 30, 1975, the Treasury Department blocked Vietnamese assets estimated at $70 million. The assets controls affected South Vietnamese dollar accounts in the United States, U.S. dollar accounts in other countries in United States or foreign financial institutions, and foreign currency accounts in American banks in the United States or abroad. Then, on May 16, the Commerce Department imposed the most stringent controls possible on exports to South Vietnam. The South, like the North, was considered a "Category Z" country, requiring all U.S. exports to Vietnam to be licensed. These controls are stricter than those applied to the Soviet Union or Cuba.

In congressional testimony, Robert Miller, then deputy assistant secretary of state for East Asian and Pacific affairs, noted, "At the present time, we deny applications for all but charitable humanitarian exports."[53] It is a policy which has continued to the present, with even some humanitarian aid encountering roadblocks.

Miller justified the assets controls as a means of protecting U.S. investments in southern Vietnam. These, he said, "could run as high as $110 million." Assets controls were imposed under the Trading with the Enemy Act. The trade restrictions were applied on the basis of the Export Administration Act's provisions which allow such controls for national security reasons. Questioned about the national security considerations involved in the case of Vietnam, where the war had ended, Miller said, "The new regimes in South Vietnam and in Cambodia came to power through force of arms against the governments that we were supporting and aiding and their policies and actions were therefore totally hostile to the United States."[54]

There were quick protests against the embargo. In a May 21 editorial, the *Los Angeles Times* said: "If the United States wants to communicate its own self-confidence, its reliability as an ally, it should get on with the normalization of relations with all nations, including the small and desperate. To pour truck factories and computers into the Soviet Union while denying all trade with Vietnam is only to invite doubts about American leadership."

In July, Louis Saubolle, head of Bank of America's Asia Representative Office, visited Hanoi. On his return to Hong Kong, he said Vietnam was eager to trade with U.S. companies and was willing to develop trade ties even though there were no diplomatic ties, in the way that U.S.-China relations had developed.[55] Saubolle drafted a resolution passed in November by the Asia Pacific Council of American Chambers of Commerce calling on the United States and the governments of Indochina to "remove obstacles" to normal trade. Robert Goodwin, vice president of the Hong Kong chapter, said the businessmen planned support for congressional efforts to lift the trade ban. "Everyone agreed this fight is something we should do," he said. "The embargo against China accomplished nothing."[56]

In October, Rep. Jonathan Bingham noted in Congressional debate that, up to that time, "the Governments of Vietnam have refrained from expropriating . . . American property."[57] The *Wall Street Journal* reported from Hong Kong November 13 that businessmen "aren't discounting the possibility that Vietnam could become an economic power to be reckoned with. There are rich agricultural areas and timber in the South and coal in the North. And there is the possibility of offshore oil." Saubolle told the *Journal* that "before too long, Vietnam could emerge as a serious competitor in the world export market."

The following March, Saubolle's office began publication of a monthly newsletter for businessmen following economic and political developments in Vietnam, Laos and Kampuchea. The newsletter, which continued for more than three years, indicated the seriousness of business hopes for trade with and investment in Vietnam.

Yet those hopes were to be frustrated. Rep. Bingham's remarks opening the June 4, 1975, hearings proved all too accurate:

> ... It has been my hope ... that our peacetime policies toward Indochina would not be mere extensions of our wartime sanctions—that the end of the fighting in Vietnam and the end of U.S. military involvement there would make possible a gradual normalization of relations with the people and governments of Indochina. Imposition of an embargo, even before the policies and new governments of South Vietnam and Cambodia have been tested, does not bode well for such a gradual normalization. Embargoes—as our experience with Cuba well illustrates—have little effect other than to prolong hostility. Rather than depriving foreign governments of needed goods, they deprive us of dialogue and influence with those governments. Symbolic gestures with little practical impact when they are invoked, embargoes often become serious hurdles indeed when the time comes for them to be revoked.[58]

Nine years later, the United States remains cut off from Vietnam and Kampuchea. There are no diplomatic relations with those countries, and the same laws which provide for the trade embargo have been used to delay licenses to private voluntary organizations attempting to respond to the basic human needs of the peoples of both Vietnam and Kampuchea.

Meanwhile, the shifting political context in Asia had its impact on the business community. The American Chamber of Commerce in Hong Kong has continued to show some interest in Vietnam, but most of the participants in Vietnam-related activities dropped out by early 1981. There were two main reasons: There was no trade between Vietnam and the United States, largely because of the embargo; and China let it be known that companies trading with Vietnam might jeopardize their recently-developed relations with China.

REUNIFICATION

"Nothing is more precious than independence and freedom," was surely the most often quoted sentence from Ho Chi Minh, the leader of Vietnam's revolution from colonial days until his death in 1969. But a second sentence was quoted nearly as often: "The country of Vietnam is one; the people of Vietnam are one." In his New Year poem to the Vietnamese people in 1968, Ho wrote, "North and South reunited, What spring could be happier?" There was never a question in the mind of Vietnam's revolutionary leaders about whether Vietnam would be reunited. The only questions were when and how.

At the end of the war, western observers generally supposed it would take two to five years before North and South would be formally united—a line of speculation reinforced by PRG officials who emphasized "certain realities and specific characteristics" of the two zones.[59] Two Quaker workers, who left at the end of July 1975, reported that the stress had been on various measures of *de facto* reunification such as letters and family visits between North and South.[60] "The basic PRG and progressive attitude toward reunification was that Vietnam had always been one country and in spirit the people were reunified with liberation," they reported. But one sign they saw indicating formal reunification might take several years was the revolutionaries' efforts "not to worry people with an abrupt rupture with the past. . . . Many people in Saigon are scared of reunification because it would mean to them that finally the communists have taken over."

By November, though, a decision had been made to move quickly to a single government of all Vietnam. Part of the reason was international. Wilfred Burchett, an Australian journalist with close ties to the Vietnamese leadership, wrote that it had seemed Vietnam could have stronger, more varied ties with the outside world if it remained divided.[61] The leaders thought it would be valuable for Vietnam to have two voices in the United Nations. And the South was a member of the nonaligned movement, while the North enjoyed close ties with the Soviet bloc. Two things changed. The North was admitted to the nonaligned movement, and the United States vetoed the admission of the two Vietnams to the United Nations, despite a vote of 123-0 in the General Assembly asking for favorable reconsideration in the Security Council.[62]

In early November, a joint commission with 25 representatives each from North and South was set up to work out procedures for reunification. The southern delegation included six members who had played prominent roles in the "third force" movements in Saigon.[63] The joint commission called for elections to be held in the first half of 1976 for a unified National Assembly. Those elections were held April 25, 1976, less than a year after the end of the war. The National Assembly gathered in Hanoi in late June. When the session ended July 3, it had declared Vietnam to be one country, the Socialist Republic of Vietnam, had named Hanoi the capital, and had given the name Ho Chi Minh City to Saigon, Cholon and the surrounding administrative district.[64] The 1959 constitution of the Democratic Republic of Vietnam would be used until a new constitution could be written.

The Chaos in the South

Official announcements indicated that the Vietnamese leaders also hoped reunification would help them get a better grip on the chaotic southern economy. ". . . The quicker national reunification is achieved," Saigon Radio said, "the sooner it will create conditions for the South Vietnam market to stamp out the vestiges of decayed and moribund economy."[65] While party leader Truong Chinh stressed recognition of the "specific conditions of the South," official statements emphasized benefits of bringing the South into the framework of the national plan. The fact that the next five-year plan was to begin in 1976 may have added urgency to this part of the argument.

The final eight months of 1975 were taken up primarily with efforts by the PRG to consolidate their control and to alleviate the most serious problems:

- large numbers of dependent populations—widows, orphans, refugees, war casualties—crowded into cities and rural camps;
- vast areas of land made fallow by bombing, defoliation, lack of labor and the resulting decline in food production;
- destroyed infrastructure, including roads, schools, hospitals, factories and homes;
- shortages of trained personnel;
- serious inflation coupled with the withdrawal of U.S. support for the economy of the South;
- the fear and mistrust on the part of many residents of the South, particularly Saigon, of the new government.

It was a set of problems Hanoi and PRG had not expected to have to tackle in 1975. Military preparations for the offensive had been thorough. Ample amounts of war supplies had been moved south. The road network known as the Ho Chi Minh Trail had been expanded and improved, and it now included an oil pipeline reaching deep into the South. But when the Lao Dong Party's Political Bureau met for nearly three weeks in December 1974 and January 1975, they really expected that the 1975 fighting would be only a prelude to the final push.[66] They were by then sure that the Thieu regime in Saigon was not willing to share power with the PRG and the non-aligned "third force"—the formula prescribed in the Paris peace agreement. They also believed the United States would not send combat troops back to Vietnam, and that without direct U.S.

intervention Thieu's forces could be defeated. They did not antici- pate how quickly Thieu's million-man army would disintegrate. They believed the final battles would not come until 1976.

Would the transition have gone more smoothly if there had been another year to prepare? It is a question which can never be answered with certainty. The revolutionaries had lost many of their most talented leaders over the course of the long war. Many were killed after they surfaced to take a public role in leading the 1968 Tet Offensive. Others died in other battles or were lost to various counter-terror campaigns, the most famous of which was the Phoenix Program. One Vietnamese spokesman estimates that as many as 90 percent of the native southern cadre may have been killed.[67] More time for study and preparation might have improved the way the new government dealt with some of the immediate material problems of the population. But another year probably would not have dislodged the assumption that the millions who had been forced from the countryside into the cities by the war would rush to return to the countryside.

The Return to the Land

The logic of encouraging such reverse migration was inescapa- ble, and the notion was not unique to the PRG. American aid officials had discussed it at least as early as 1970.[68] Both Saigon and the revolutionaries had encouraged resettlement of rural areas they controlled during the two-year period between the signing of the peace agreement and the end of the war. When the war ended, this became an urgent priority. "The day American aid stopped," a foreign diplomat commented, "Saigon couldn't sustain itself. There was a surplus of at least two million people."[69]

Out in the provinces, the return to the land appeared to be swift, spontaneous and widespread. An American relief worker who remained in the central provincial town of Quang Ngai saw entire refugee camps vanish in a matter of days.[70] ". . . [A] man was swinging a heavy club, smashing the mud walls of his refugee house. Others were pulling the thatch off the roofs of other huts," he reported. "In fact, most of the mud buildings in the camp were already demolished. Men were salvaging the bamboo posts that had formed the framework of the houses. Others were taking beds, tables, cooking pots, bamboo poles or a few sheets of USAID- supplied tin roofing and tying them onto bicycles or shoulder poles. Already the road ahead of us was streaming with people pushing their loaded bicycles or jogging to the rhythm of swinging shoulder poles."

What they found when they returned to their old homes was

often profoundly discouraging. So little was left of these often fought-over villages that people could no longer determine whose house once was where. Along the coast, dikes had been destroyed by the fighting. Once the sea could be drained from their land, it would still take some years before enough salt could be washed from the soil to allow rice to grow again. Still, these people were used to the demanding life of one of the country's harsher provinces. Quang Ngai had suffered relatively little from the "development" of the commercial sector which had catered to foreign military bases in larger cities. Relatively few of the war refugees had become accustomed to the false prosperity induced by the flow of American dollars in Saigon and Danang. Quang Ngai was not so badly dislocated by the war in some important ways—in the motivations of its people and the values woven into the social fabric—as places like Saigon which had been relatively immune from the direct impact of combat.

Return to the land from Saigon was more structured. In Saigon as in other towns and cities, "military management committees" had been given administrative authority until neighborhoods, wards, and finally the whole town could organize "people's committees" to take charge. The neighborhood committees took charge of registering people to return to the land and arranged transportation along with a sum of money or rice to tide them over during the trip.[71] In the first month, about 18,000 people left Saigon and somewhere near three times that number had requested to leave. In some cases, families would send one member to test the water.[72] If things looked good, the rest of the family would follow. High government officials were overly optimistic about the prospects for this program to repopulate the countryside. Minister of State Nguyen Van Hieu said, shortly after the end of the war, that the country's rice fields could immediately absorb more than a million people.[73]

But Vietnam's fields were dangerous and the work of restoring them to production was arduous. Settlers who could not be welcomed back to fields already under cultivation had to deal with problems ranging from soil compaction to invasion by persistent weeds to the presence of unexploded mines and grenades. The unexploded munitions were often difficult to detect. Some, like the M-79 grenade, were barely larger than an egg; struck by a farmer's hoe, however, they could claim an arm, a leg or a life. They were a major problem for years after the war ended, and the countryside is still littered with this explosive debris. Considering the fact that munitions from World Wars I and II are still found, unexploded on European and Asian battlefields, and the fact that the United States rained an unprecedented seven million tons of ordnance on Vietnam (to which must be added the much smaller amount of munitions used by the revolutionary troops), this will likely be a problem for Vietnam for many years.

Later reports made it clear that the pace of rural resettlement slowed, and many settlers returned to Saigon, often to eke out a livelihood on the fringes of society.

Attempts to Control Southern Economy

Within Saigon and other towns and cities, the new authorities faced unemployment variously estimated at two to three million, including a million former members of the Thieu government's army. Saigon's poor, according to the official city daily *Saigon Giai Phong*, were menaced by hunger.[74] And the revolutionaries feared that traders, bankers and factory owners might sabotage the economy.

Despite looting of official stocks of rice during the transition period, the government was able to provide some rice for free distribution. Between May and December, these emergency distributions had totaled 135,000 tons.[75] The country's banks, which had ceased to function a few days before PRG troops arrived, were kept closed for more than two months while PRG cadre examined the accounts.[76] Assets of some foreign firms were also scrutinized. The inspectors found that nearly half the money supply of the South, worth about $200 million, had been withdrawn from the banks during the final month before the old regime collapsed.[77] For months, businessmen continued to hoard their cash, rather than invest it. When the banks did reopen, essentially as branches of the newly-formed National Bank of South Vietnam, there were strict limits on withdrawals. This had the ironic effect of making it difficult or even impossible for factory owners to pay their workers or buy raw materials, though the government was exhorting them to join in helping to "build an independent, self-contained economy."[78] Still, much of the city's industry was in operation, limited primarily by shortages of fuel and previously-imported raw materials.[79]

Then in September came the government's first decisive attempts to gain control of the economy. The most sweeping measures came September 22, 1975. In an announcement which took most of Saigon by surprise, the government banned unnecessary movement on the streets of the city and gave people five hours to register their accounts and trade in their old Saigon piasters, which had remained in circulation until then, for banknotes of the Bank of Vietnam.[80]

The money exchange followed by less than two weeks the start of a campaign against the city's most prominent speculators in commodities ranging from rice and cloth to barbed wire.[81] The September 11 raids on homes and warehouses of the largest wholesalers were the first in a series of measures, still continuing with limited success, to establish a state trading network which can supplant the private market.

Attempts to Control Foreign Aid

As it was taking these internal measures, the PRG was also attempting to gain control of U.S. aid shipments which had been on their way to Vietnam at the time of the PRG victory.[82] The shipments included rice, wheat, dried milk and blankets which would have been useful for the government's relief and resettlement efforts and machinery, spare parts and industrial chemicals which would have helped keep the country's industries in operation. Total value of the shipments was estimated at $50 million.[83] American officials, operating under terms of the trade embargo, which had just been extended to South Vietnam, prevented any transfer to Vietnam and auctioned off the goods.

There were some brighter developments in international relations. Japan, for example, agreed in October to provide nearly $28 million in grant aid to the North.[84] Both China and the Soviet Union continued to provide aid, though some strains with China were already evident. American oilmen responded with some interest to Vietnamese suggestions that they might be able to resume their explorations. They were, however, blocked by the trade embargo. "At the present time our problem rests in the lap of the gods in Washington," a Cities Service executive said. "Clearly a change of heart is needed there before we can do anything further in Vietnam."[85]

U.S.-Vietnam relations remained at a standoff. In November, Nguyen Thi Binh, the PRG foreign minister, said that normalization of relations between the two countries depended on Washington's attitude.[86] "Our policies toward the new regimes —in Indochina," President Ford countered, "will be determined by their conduct toward us. We are prepared to reciprocate goodwill—particularly the return of the remains of Americans killed or missing in action, or information about them."[87] There were no suggestions, however, that the United States would meet what Vietnamese leaders called its "obligations in the healing of the wounds of war"—a point the Vietnamese continued to insist be met.[88]

Opening of Railroad Symbolic Link

Reopening the rail link between Saigon and Hanoi was also a high priority for the PRG when it took power in South Vietnam on April 30, 1975. The rail link was certainly important for transport in the long, narrow country, particularly for transporting rice to the North and fertilizer to the South. But it was at least equally important as a symbol of reunification in a country divided for 30 years by war and politics.

The rail system in Vietnam, developed during the colonial era, was not extensive. The main line linked Hanoi with Saigon. From Hanoi, the line continued northwest to Lao Cay and on into China. Another branch reached up nearly to the China border in the northeast, passing through Lang Son, then turning to the northwest for a few more miles. A third branch from Hanoi ran east to the port city of Haiphong. In the South, one branch ran up to the mountain resort of Dalat. Another connected Saigon with the town of Loc Ninh in the heart of the rubber plantations which sprawled across the Vietnam-Kampuchea border. A southern spur reached to My Tho, the first major town in the Mekong Delta. No rail lines connected Vietnam with the other former French colonies of Kampuchea and Laos.

Rail lines are always important targets in war, and in Vietnam there was no exception to this rule. In the South, National Liberation Front saboteurs knocked out rail bridges and sections of track with mines and demolition charges. Although there were occasional attempts to restore parts of the railroad, these were largely undertaken as symbolic demonstrations that the area involved had been "pacified." Because of the huge American air transport capability, rail lines were not crucial to the war effort. Civilian traffic often used buses and trucks, but there was also a civilian air line far more extensive than could otherwise have been justified in a country with such a weak economy. In deserted areas, railroad track often disappeared to be sold as scrap or used for reinforcing bunkers.

In the North the rail line—especially bridges and switching yards—provided major targets for the American air war. Because the North could not afford the alternate means of transport used in the South, and also as a matter of pride, great efforts went into keeping the rail line open, though there was surely a deterioration in the quality of roadbed and rails during the war, and many of the bridges were not permanently replaced until after the war.

Thus it was a major announcement in late November when PRG officials declared the first 426-mile segment reaching north from Saigon to the small town of Phu My, between the provincial capitals of Qui Nhon and Quang Ngai, was open for traffic.[89] Work was continuing with the goal of opening the entire 1,050-mile line by mid-1976 to be in operation to greet the National Assembly meeting. Despite this earlier optimism, the symbolic rail link between Hanoi and Saigon was not operating in April. It was a year later—June 20,

1977—before regular passenger service resumed on the line. That slippage between hopes and reality was to be repeated in other, more crucial areas of economic development.

THE FIRST FIVE YEAR PLAN

Vietnam's first five year plan period since the end of the war began in 1976, although the plan was not actually approved by a congress of the party until the end of 1976—one indication of disagreements within the party leadership on what economic policies the country should follow.

Differing Perspectives

The discussions within the Political Bureau may well have continued debates which some observers said had begun in the summer of 1975—a debate on how quickly the South should be moved toward socialism.[90] Prime Minister Pham Van Dong was said to be on one side, arguing for a long, gradual transition period. Defense Minister Vo Nguyen Giap and National Assembly Chairman Truong Chinh were said to be arguing for a faster transition. Another theme, likewise crucial to the shape of the five-year plan, was surely the question of how much emphasis to place on heavy industry as compared to light industry and agriculture. In one form or another, it is a question which faces any Third World country which seeks economic independence. In socialist countries, it is also a matter of doctrine. A book review in the April 1975 issue of the party theoretical journal *Hoc Tap* stressed the theme that a socialist society needed heavy industry to provide the "material foundations" of socialism. The central problem, the article argued, was industrialization, particularly in the North. This is an approach frequently associated with the party's general secretary, Le Duan. In economic development policy, it stresses building heavy industry as rapidly as possible so that the country can produce its own farm machinery, construction materials, fertilizers, and machinery to equip the light industry which, in turn, supplies consumer goods. In human relations, this approach argues that a socialist society cannot finally be built until there are enough goods to go around to provide a comfortable life for everyone. Both of these points are, of course, oversimplified, but they capture essential elements of one tendency in a debate which still continues in Vietnam.

The other tendency stresses the fact that Vietnam somehow must pay for its economic development. In order to earn the foreign exchange to pay for industrial development, indeed to pay for fertilizers and pesticides for agriculture in the short term, Vietnam must export something. Much of the country's present export capacity is in areas such as textiles and processed foods (light industry), handicrafts and such agricultural cash crops as tea, coffee, fresh fruit and fresh vegetables. Those who stress this point of view call for a delay in investment in heavy industry with current investment going mainly to agriculture and light industry. Once these sectors are healthy and earning reliable foreign currency income, this argument goes, then Vietnam will be in a position to move ahead with heavy industry.

Despite differences in emphasis, both sides agreed that industry should support agriculture at this stage in Vietnam's development. Electricity is needed, for example, to power pumps for irrigation and drainage of fields. Cement is needed to build drying yards for rice and other staples. Petroleum production would give a domestic source of fuel and nitrogen fertilizers. Increased production of consumer goods can offer farmers something to buy with their money if they increase production.

In debates over human relations, the division must surely follow different lines. A main theme is that people can be inspired to change their social structures, to become "new men and women," even before there is a material base strong enough to provide a comfortable life for all. Such arguments are frequently associated with Truong Chinh. (He and Le Duan are the party's two major theoreticians.) Vietnam's wartime experiences seemed to validate this "inspirational" line, but postwar experience has dealt it heavy blows.

The Plan Revealed

Whatever the debates of the Political Bureau and the Central Committee, the five-year plan presented to the party congress was ambitious. By the end of the plan, deficits in staple foods were to be eliminated. The target for 1980 was 18 million tons of paddy (unhulled rice, which yields 60-65 percent of its weight in finished rice) plus other staples equivalent to three million tons of paddy.[91] This "leaping advance of agriculture," as the resolution of the party congress called it, was to provide enough for domestic food and animal fodder plus a small surplus for export. Other goals included clearing a million hectares of new farmland and increasing the pig population to more than 16 million (compared to 8.8 million in 1975).[92]

"In 1976," Agriculture Minister Vo Thuc Dong explained these goals, "Vietnam produced about 12.5 million tons of paddy; so from now to 1980, an increase of more than 5 million tons will be necessary, or an average yearly increase of 8 percent. This is a fairly rapid but entirely feasible tempo of development." Subsidiary crops like corn, manioc, potatoes and sorghum would have to be increased by nearly four times. Soybeans, peanuts, sesame, and coconuts were to be promoted "in order to increase the protein and fat content of people's diets." And to provide raw materials for industry and cash crops for export, the country needed to find areas for growing cotton and step up production of rubber, coffee, cocoa, tea, sugar cane, pineapples, and bananas.

Pig raising, Dong said, was important not only for meat but also as the "main source of organic fertilizers, a substantial contribution to increasing the output of food and other crops." Besides nearly doubling the number of pigs being raised by families, collective groups and state farms, Dong predicted an increase in the average weight of the animals.

Targets for industrial growth were nearly twice as large as those for agriculture. Gross industrial output was to rise 16 to 18 percent a year. The Central Committee's report to the party congress called for "great efforts to build a number of new heavy industry establishments, especially of the engineering industry, as soon as possible."[93] The engineering industry, the report said, "must become the prime concern in the whole plan for the development of the national economy." By the end of the plan, Vietnam was to be producing high quality machine tools to meet needs for spare parts, "turn out large quantities of machine tools, tractors, sea-going ships, trawlers, dredgers, hydroelectric turbines, etc." Energy industries were to be developed "rapidly in anticipation of the needs of the national economy," including first steps toward building a unified power grid for the whole country. Steel capacity was to be expanded and other metals—tin, lead, zinc, copper and aluminum—added to the country's production "in a way commensurate with the development of engineering, construction and other branches." The emphasis in the chemical industry was to be fertilizers and pesticides. And construction of plants to make synthetic fibers, caustic soda and other industrial chemicals would need to begin "soon."

Investment of money would favor industry, with 35 percent of total investment allocated for industry and 30 percent for agriculture. Considering that industry produced less than a quarter of the national income in 1975 while agriculture produced nearly half, the tilt toward industry seems even more pronounced.[94] In fact, investment in agriculture was considerably below the plan's projections, averaging less than 26 percent for the five-year period.[95] Actual industrial investment was close to the projected 35 percent, but that was 35 percent of a smaller total. The total invested in all sectors turned out to be only 17.8 billion *dong* rather than the anticipated 30 billion. This was not the only way in which reality fell short of the plan's projections. At times it must have seemed that every power from the heavens to the bedrooms had conspired to make life difficult. A population survey in 1976—part of the preparation for the National Assembly elections—showed 48.8 million people in the country, over four million more than 1974 estimates had indicated.[96] This was partly an indication of the inaccuracy of earlier estimates. But it also reflected a population growth rate of 2.6 percent, which added a million mouths to the number which needed to be fed each year. The growth rate was confirmed by the formal census in October 1980, which put the country's population at 52.7 million.

Even before the plan period began, Vietnam had a pledge from the Soviet Union to provide training and investment in industry and agriculture. The aid package included 40 capital projects, among them a massive hydroelectric project on the Black River in the northern delta, a thermal power plant, a coal mine and a caustic soda plant.[97] Western diplomats estimated that value of the package at perhaps $500 million. China pledged help with a steel mill, a fertilizer factory, a new bridge across the Red River at Hanoi, and a continued supply of less expensive but vital items like consumer goods and pharmaceuticals.[98]

Hopes for Western Investment

Vietnamese planners were also counting on substantial amounts of aid from Japan and the West. "If they are going to import technology, as they must, they will insist on the best," the *Far Eastern Economic Review*'s Malcolm Salmon noted in late 1975.[99] "Some Vietnamese friends have indirectly criticized the North Koreans, for example, for over-reliance on Soviet models in factory design, which in many instances have turned out to be not up to scratch in the industrial world of the 1970s. Of their intended connections with capitalist concerns they say: 'We're not afraid of losing our independence. If half a million GIs couldn't subjugate us, a few dozen foreign companies aren't going to.'"

Vietnamese officials were particularly careful to encourage good relations with Japanese companies which had opened factories in the South before the end of the war. Though the Japanese managers had been evacuated before the Saigon regime collapsed, the new government did not nationalize the factories, which were assembling radios, televisions, pumps and tillers from parts manufactured in Japan.[100] Many French plant managers remained in the country, and were allowed to continue running their businesses under the supervision of workers' committees. French-owned enterprises included a shipyard, a bicycle factory and a brewery. The Vietnamese goal was to turn these industries into joint ventures, with Vietnam holding majority ownership in the resulting companies.

In April 1977, Vietnam announced its foreign investment code, which allowed three forms of investment:[101]

- "Cooperation in production" in which the foreign investor would receive profits in kind. In most cases. the investor would be required to export its share of the products.
- Joint enterprises. Both of the first two types of enterprise envision use of local raw materials. In the case of joint enterprises, production would be for export, and profits would be divided in cash rather than in goods.
- Production exclusively for export. These firms would be "export processing" factories, importing raw materials and exporting finished products.

The final version of the code reflected a Vietnamese willingness to yield to at least some of the complaints potential investors registered when a preliminary draft was circulated several months earlier.[102]

One of the few areas where foreign firms actively sought to invest in Vietnam was in oil explorations. Mobil and Shell, which had achieved the most promising results in early-1975 exploration off the southern coast, would have been in the best position to move ahead. Mobil's find—one well which tested at a rate of 2,400 barrels a day of light crude—was the best.[103] Mobil plugged it and took the "treasure map" with it when it halted operations in 1975. Under the terms of the U.S. trade embargo, Mobil and Shell were both blocked from joining the exploration. Not only that, but non-American companies were prevented from using U.S. technology in exploration or production. Offshore oil and gas operations are one area where American technology is the clear industry leader, so the embargo was a serious embargo to any western participation.

By September 1977, Vietnam had signed agreements with a number of European companies to explore for oil off the southern coast.[104] A Canadian firm later joined them in what turned out to be fruitless searches.[105] The explorations found some traces of hydrocarbons and other promising geological structures but nothing with commercial potential. The operations proved more costly than the companies had anticipated, and they pressed their Vietnamese partners for more favorable terms but were turned down. None of these companies was exploring the block where Mobil had made its find in 1975. This was one of several areas Vietnam reserved for its own exploration, and it was expected that joint Vietnamese-Soviet exploration would take place there. Although prospects for oil have frequently been cited by Vietnamese leaders as a cause for hope in the country's economic future, there have been no reports of a strike by Soviet crews.

The U.S. trade embargo, however, was not the only obstacle to foreign investment. Nayan Chanda, a close observer of Vietnam's economic development efforts, noted that "Vietnam's otherwise liberal foreign investment code failed to attract more than a handful of Western investors" in important part because of the "deep, almost xenophobic suspicion about the motives of those who want to invest in the country. This, coupled with the slowly-grinding bureaucratic machine and indecision have discouraged potential Western investors."[106]

If hopes for western investment seemed stalled, there had at least been agreement on grants and loans to finance a number of significant projects.[107] France and Sweden offered the largest amounts. The state-of-the-art Swedish paper mill project included development of pulpwood forests as well as construction of the mill itself. Danish aid included a high-technology cement plant, which provided an interesting balance to a Soviet cement plant being built on a similar schedule. Other donors included Finland, Japan and India. And both the International Monetary Fund and the Asian Development Bank approved loans.

There were encouraging signs, too, on the possibility of improved relations with the United States—at least in the first part of the administration of President Jimmy Carter. Carter recognized that the issue of Americans missing in action in Indochina was a volatile one. So in early 1977 he sent a commission headed by Leonard Woodcock to Hanoi to investigate. The group found a "spirit of goodwill" in Hanoi.[108] About the same time, the Carter administration symbolically relaxed some provisions of the trade embargo.[109] The Woodcock mission led to the resumption of direct U.S.-Vietnam talks on normalization—talks which eventually stalled over the issue of U.S. reconstruction assistance.[110] By July 1978, when Vietnam publicly dropped its insistence that the United States agree to assist Vietnam before diplomatic relations were resumed, the administration was moving rapidly to full diplomatic relations with China.[111] Vietnam and its former ally in Kampuchea, the Khmer Rouge, were on the brink of war. And China, which had cancelled its aid to Vietnam in May, was increasingly hostile to a Vietnam it saw as a Soviet proxy in Southeast Asia. This unfortunate convergence of unexpected events on the international scene was further complicated by developments inside Vietnam.

Weaknesses in the Plan

In 1976, when the weather was especially favorable it seemed that agriculture would be capable of the "leaping advance" called for in the state economic plan. Total production of rice and other staples was 13.6 million tons, more than 17 percent above the 1975 figure of 11.6 million.[112] Both the area planted to industrial crops and the yield of most of those crops also increased. Sugar cane output rose 67 percent, soybeans, 47 percent, and peanuts, nearly 40 percent. Coffee production was up more than 40 percent on reduced area. Jute, rush, tobacco, tea, rubber and coconuts showed more modest increase. Total agricultural production was up 10 percent—at the upper end of the targets set in the plan.

Then, rice production fell sharply in 1977. In 1978 slumped below 1975 levels. Only the large, steady increases in secondary staples kept total staple production in the 12.9 million ton range both years. Production did recover to 13.8 million tons in 1979, slightly more than 1976 output, but of course there had been a population increase of about three million in the intervening years.[113]

Much of the drop in production could be blamed on the weather. Conditions in 1976 had been especially good. Then Vietnam's crops were hit by unusually harsh conditions for four years in a row, including both droughts and floods in the same year.[114] One year a major problem was too many cloudy days—not enough sun for proper maturing of the crops.

This was also a period of building tensions with Pol Pot's Khmer Rouge government of Democratic Kampuchea. Many areas along the Vietnam-Kampuchea border had been scenes of heavy fighting during the American war, and thus were the sort of abandoned lands Vietnamese planners were turning into new economic zones as part of their effort to reduce urban population and increase the area planted in crops. They hoped to relocate hundreds of thousands of people to the border provinces, bringing settlers both from Ho Chi Minh City and from crowded rural provinces of the North.[115] These plans were seriously threatened when Kampuchean troops mounted raids on border villages beginning in 1977. Thousands of settlers fled the area, some of them even camping on the streets in Ho Chi Minh City. Plans for new economic zones in the border region had to be suspended.

Another burden on Vietnam's strained economy was the influx of refugees fleeing the Pol Pot government in Kampuchea. By the time of the Vietnamese invasion, there were some 160,000 people from Kampuchea in refugee camps in Vietnam.[116]

Other problems reflected weaknesses in Vietnamese plans. In the North, there had been a drive since 1975 to encourage existing cooperatives, which normally included a single village with 200 to 300 hectares of land, to join together in twos or threes to form higher-level cooperatives.[117] But these larger co-ops, sometimes encompassing up to 1,000 hectares, were too unwieldy for the management skills available.[118] The organization of work was frequently inefficient, with work crews wasting hours travelling to one of the more distant corners of the co-op. Production stagnated or even fell. Besides management problems, there were shortages of inputs such as fertilizer which the government was supposed to supply to farmers. One Vietnamese critique noted that during some crop seasons the supply of nitrogen fertilizers had dropped to half or even a third of what had been supplied in previous years.[119]

The border clashes with Kampuchea continued. In December 1978, Vietnamese troops, together with former Khmer Rouge who had revolted against the Pol Pot regime and requested Vietnamese support in ousting it, swept into Kampuchea and deposed the Pol Pot government. The invasion force was generally estimated at 120,000. The number of troops assigned to Kampuchea later rose as high as 200,000, according to western estimates. Now, about 160,000 Vietnamese troops are said to be there.

The following February, China retaliated with a punitive expedition into Vietnam's northern border area. Continuing tensions on the Vietnan-China border led Vietnam to station an estimated 250,000 troops there.[120] This sudden mobilization brought a drop in industrial production, and altered priorities. One example reported by Nayan Chanda: "Mai Dong engineering works has not only sent 35 percent of its male labor force to the army, but has had to switch from producing metal presses to making shovels—urgently needed to dig trenches."[121] Troops which had been raising crops and clearing land were placed on combat duty. And to keep the 1978 deficit within bounds, capital expenditures were cut 30 percent.

It was also in this period that Ho Chi Minh City authorities mounted their March 1978 campaign intended to abolish "bourgeois trade."[122] It was the most dramatic attempt of that time to break private control over trade. Because the state trade machinery was still too weak to take over from the private traders, the drive had limited success. It did cause consternation among those ethnic Chinese who made up the majority of the city's wholesalers. China assailed Vietnam for persecuting its Chinese residents. Vietnam accused China of spreading rumors among the ethnic Chinese community and of instigating the departure of ethnic Chinese from Vietnam.[123] Within two months, more than 100,000 ethnic Chinese had crossed Vietnam's northern border into China. It was this influx—the need to care for so many new people—which China cited in its May announcement that it was halting aid projects in Vietnam. The value of the canceled projects was estimated at $500 million. Most of the refugees crossing the border into China were from northern Vietnam. There was a corresponding rise in the number of ethnic Chinese among the boat people arriving in other Southeast Asian countries. Refugee arrivals doubled between July and October, and three-quarters of those refugees were ethnic Chinese.

A second wave of ethnic Chinese left after the February 1979 Chinese attack on Vietnam. They said they had been offered a choice between moving to new economic zones distant from the China border or leaving the country. Whatever the merits of Vietnam's "national security" concerns, this second wave of ethnic Chinese departures hit the economy in vital areas. Some 3,000 of Hanoi's ethnic Chinese, for example, worked in government offices.[124] Hospitals, schools and research institutes were also depleted, as were a number of skilled trades. More than 15 percent of Vietnam's coal miners were ethnic Chinese, and ethnic Chinese were well represented in the work force of Haiphong, the main northern port. Soviet longshoremen were brought in to untangle port congestion. Coal production sank in 1979 and 1980, halving exports of the country's single most important source of hard currency.[125]

With little more than a year to run on the five-year plan, an assessment in the party daily *Nhan Dan* outlined the "great economic difficulties affecting our people's lives. These difficulties consist of lack of food for the people, raw materials for factories and fertilizer for rice paddies; insufficient consumer goods; and too little foreign exchange to pay our creditors."[126] Still, the commentary insisted, "Our economic situation is not yet favorable, but it is not bleak."

It was about this same time — in October 1979 — that the government began publicizing a sweeping package of reforms adopted by the party's Central Committee. Because the resolutions were adopted at the sixth plenum — the sixth meeting of the full Central Committee since the 1976 party congress — they were tagged Resolution Six. They called for more private initiative, more regional autonomy, and a shift in emphasis toward production of consumer goods.

On the international scene, the Vietnamese troops in Kampuchea prompted a negative reaction from some countries which had been providing aid to Vietnam. Japan, for one, suspended delivery of $56 million it had promised in late 1978.[127] Australia also suspended aid. The United States was very public in its criticisms of the Vietnamese occupation of Kampuchea, and in at least one reported instance used those criticisms to try to block U.N. Development Program (UNDP) assistance.[128] The American Congress was also hostile. In September 1979, the House of Representatives voted to cut funding for the World Bank, the Asian Development Bank and other international financial institutions, and to condition American contributions on a provision barring use of U.S. money for loans to Vietnam.[129]

The international institutions insisted they could not accept funds with such conditions attached. To break the impasse, World Bank President Robert McNamara, the former U.S. defense secretary, assured the Congress that the World Bank would provide no loans to Vietnam in fiscal 1980.

THE BOAT PEOPLE

Ethnic Chinese

Ethnic Chinese leaving Vietnam in 1978 and 1979 fall, in large part, into two relatively distinct groups. The first includes traders who decided to leave after the March 1978 crackdown on "bourgeois trade." The government mobilized a combined force of soldiers and children to inventory every business in the city.[130] All private wholesalers were ordered to shut down. Small businesses were allowed to continue operating legally. Officials have said that the measures were applied impartially to Vietnamese and ethnic Chinese alike.

For several reasons, the government moves against private trade were of special concern to the ethnic Chinese. Most important was the fact that they controlled some 80 percent of commercial activity in the city.[131] (It should be noted, however, that most of the million ethnic Chinese in the South were not wholesalers or even entrepreneurs. The majority were ordinary workers.) The growing tensions between Vietnam and China added to the unease in the Chinese community. And it may also have been significant that the size of the Vietnamese market was shrinking. There was less money to be made from trade. This combination of reasons impelled several tens of thousands of ethnic Chinese to join the exodus of boat people from southern Vietnam.

In later years, Vietnamese officials made a point of introducing foreign visitors to ethnic Chinese who had remained— businessmen as well as ordinary workers.[132] One, Tieu Nguyen Huu, said that he had switched from selling fish nets to doing construction work, refurbishing hotels and building facilities for oil exploration crews in Vung Tau. The general level of free market activity, which has remained more frenetic in old Cholon than in Saigon, indicates that the 1978 departures at most restrained the ethnic Chinese commerial networks. Five years later, they are still active.

The second major category is ethnic Chinese who were forced out of the North after the February 1979 Chinese invasion of Vietnam. Vietnamese officials harbored fears of ethnic Chinese supporting the invasion. "Many Hoa —ethnic Chinese were 'fifth columns' during China's invasion, serving as spies and advance troops and supporting China," a Vietnamese official said in May 1979.[133] "As a result, in a temporary security measure we are asking the Hoa who have maintained Chinese citizenship to go to the new agricultural areas. If they don't want to go, they are free to leave the country." These security fears and the official response reminded many observers of the American internment of Japanese-Americans during World War II. The atmosphere of wartime suspicion reinforced anti-Chinese prejudice. The results were reflected in a teenage school girl's diary:[134] "Because my mother is Chinese, there is something about me which is Chinese. As a result I have experienced such pain and anguish in my heart. In school my friends pick on me, saying, 'Since you are Chinese, why don't you go back to China.' Though I do my duties, I am confronted with so many difficulties and troubles, like having rocks thrown at me. . . ."

Among the northerners who made their way to Hong Kong were a former army major, a party member who had served with Ho Chi Minh in the Viet Bac resistance zone and with Vo Nguyen Giap at Dien Bien Phu. Another was a former planning cadre at the Ministry of Supply, ethnic Vietnamese but married to an ethnic Chinese woman. So many ethnic Chinese doctors had left that, according to a Vietnamese official interviewed in late 1979, "Now you can find parts of Vietnam where there are none."[135]

Ethnic Vietnamese

Ethnic Vietnamese who have left are a more diverse group, and it is more difficult to categorize their reasons for leaving. The cost generally reported is in the range of $1,500 to $3,000 per person.[136] Two Vietnamese-speaking Quaker workers who interviewed refugees in Malaysia in early 1979 said the one common thread in the various reasons people gave for leaving their homeland was a feeling that there was no place for them in Vietnam any more.[137] A protestant minister told them he had been arrested once on charges of connections to the CIA, held for a day, released, then questioned in his home. When he heard rumors he would be imprisoned, he decided to leave. A former lieutenant in the old Saigon army returned home from re-education in 1977 but found no work. He was not willing to become a farmer. A family from Saigon left because they could not find work. A landowner and farmer from the southern province of Ca Mau was fed up with the rice tax which, he said, left him with only enough rice for seven months of the year. To that he added the concern of conscription, and said he was not going to send his children off anywhere.

"Many were involved in business, or the profession, or had hopes of advancing in these fields, and perceive that this is not possible for

52

them in modern Vietnam," wrote the Quaker observers. "The alternatives which are open to them are often farming or physical labor, and these are not appealing." Others who have left had valuable technical skills and were given work in their fields in Vietnam. But as one petroleum engineer, now in the United States, reported, he was not trusted even after some years of loyal service under the new regime.[138] He found it unbearable to be under constant scrutiny and suspicion from a manager appointed for his political reliability rather than his technical knowledge.

The atmosphere of suspicion has certainly been nurtured by the overt hostility of American policies. And the urban unemployment would surely have been less if Vietnam had had access to American spare parts and raw materials. Probably fewer people would have left Vietnam if U.S. policy had been more positive, and it might have been easier for those who did leave to arrange legal departures—as in the current "orderly departure" program—rather than the perilous boat journeys. But life in Saigon and Cholon would still have been a comedown for those who had benefitted from the dollars which flowed so freely during the American years. On its own, Vietnam could not sustain a Honda economy; even a bicycle economy is a struggle. The bloated cities had to go on a crash diet. Even under better circumstances, there would have been some who felt there was no place for them in postwar Vietnam.

NEW ECONOMIC ZONES

Refugee accounts of life in postwar Vietnam frequently refer to new economic zones (NEZs) as a form of imprisonment at hard labor. A 1976 United Nations mission reported: "The Mission is convinced that the Viet-Namese Government's policy of establishing new economic zones meets an urgent need for which it is hard to find any other solutions."[139] Why this sharp contrast?

Life in the zones is hard. The NEZs are the frontier in a poor land. More often than not, the land is dangerous, infested with unexploded munitions. Government programs promoting the NEZs, especially in the early years, suffered from too little research and advance planning and too little investment in preparing the land for settlers. While moving to the zones is a "voluntary" decision for settlers, the "persuasion" of government cadre can sometimes seem coersive. And among the youth crews now sent to prepare new NEZs for settlers, there are at least some who are there as an alternative to other punishment for delinquency.

On the other side of the balance, in late 1982, Ho Chi Minh City still had more than 220,000 unemployed, with little immediate prospect of creating so many jobs in the industrial sector.[140] The country still needs to expand agricultural production, and there are still several million hectares of land which could be opened to cultivation.

Most western reporting focuses on Ho Chi Minh City and NEZ settlers who have come from Ho Chi Minh City. But it should be noted that the "redeployment of the labor force," as the Vietnamese call it, is a nationwide program. People are leaving from the cities of Danang and Hanoi and from the more crowded rural provinces in central Vietnam and in the northern Red River Delta. Of the 1.5 million people who moved to NEZs during the 1976-1980 plan period, only about half were from Ho Chi Minh City.[141] In fact, since most of the 300,000 said to have left the NEZs to return to their old homes were probably from Ho Chi Minh City, the city probably contributed less than half to the net settlement figure.

Some of the early horror stories from the NEZs were beyond the government's ability to control, especially the Khmer Rouge attacks on zones near the Kampuchean border. Other settlers fled the zones because of conditions created by poor government planning. Le Minh Xuan, for example, was one of the early disasters.[142] The farm is in the suburban "green belt" which is part of the Ho Chi Minh City administrative district. The first settlers were given rice seed and tools and turned loose to develop their farms. But the acid, marshy land would not support rice. Belated analysis by agricultural technicians showed that the land, if properly drained, would be excellent for pineapple—which has the advantage of being a good export corp. By 1980, adjoining marsh land was being drained for new state farms, also to be planted in pineapple. A drive from the newest "frontier" back through Le Minh Xuan—now a model NEZ— provided a graphic demonstration of the progress being made in the area. The newly opened farms were quite primitive. People lived in thatch lean-tos, each a long roof set on the ground like barracks without walls. In slightly older areas, people were building individual thatch houses, then building houses of brick. In Le Minh Xuan itself, the homes were surrounded with fruit trees and vegetable gardens.

The new regulations on NEZs issued in 1980 also stressed adequate investment.[143] A subsequent Vietnamese report noted that it costs 3,000 to 4,000 *dong* to reclaim a hectare of virgin land.[144] (At the official rate of exchange $1.00 equals about 10 *dong*.) Plans only allocated 1,200 *dong* per hectare during the 1976-1980 period, and in practice only half that amount was available. Investment at one current showcase NEZ is even higher, and includes 20,000 *dong* per family for construction of a wooden house, already in place when settlers arrive.[145]

Plowing the fields
in Ha Bac Province,
by John Spragens, Jr.

Cook at roadside
restaurant, Ca Mau,
by John Spragens, Jr.

Soldering the base
to kerosene lamp,
Quyet Tien Co-op, Cholon,
by John Spragens, Jr.

Children at a rural
child care center,
Nhu Quyen Co-op,
by John Spragens, Jr.

War-dead memorial cemetery,
Minh Hai Province,
by John Spragens, Jr.

Black market stand
in Ho Chi Minh City,
by John Spragens, Jr.

Families called out to
help bail flood waters
from the rice fields
in northern Vietnam
after typhoon,
by John Spragens, Jr.

Taking rice plants to
the fields for transplanting,
Minh Hai Province,
by John Spragens, Jr.

Bailing water from
flooded rice fields,
Ha Bac province,
by John Spragens, Jr.

A British reporter who visited this showcase, Nhi Xuan, about 30 kilometers from downtown Ho Chi Minh City, said most of the settlers were either unemployed or without "a stable way of life"—a phrase indicating people living or trading in the city without official registration papers.[146] Settlers were given basic furniture and agricultural tools in addition to the free house, and were fed by the government for the first six months. After that, they were eligible to buy up to 21 kilograms of rice per month (for adults) at official prices, which are about a tenth the price on the free market.

The land had been cleared by Assault Youth groups, and was ready to be planted in sugar cane. In theory, each family would cultivate one hectare of cane, and would be able to harvest a crop of up to 25 tons. The first 15 tons were to be sold to the government at official prices. Any surplus above that amount could be sold on the open market. Since the first crop had not yet come in for most settlers, the theoretical projection that farmers would make about as much as mid-level cadre had not yet been tested.

The reporter, fluent in Vietnamese, noted contrasting impressions from private conversations with two settlers at Nhi Xuan:

> One man in his mid-30s had spent more than seven years in a new economic zone in Hau Giang province, near the Cambodian border. At first he and his family had been able to make a living there "but gradually things got worse. Finally there was no water," he said. He came directly to Nhi Xuan, planted pineapple and sugar-cane, and so far has been making about *Dong* 1,000 a month.
>
> A neighbor seemed less satisfied. He had spent a year in another zone, he said, then went back to Ho Chi Minh City to live with relatives. His family had just finished their six months on government food and were worried about the future, especially because of insect infestation which had attacked some of their crops. He had come to Nhi Xuan, he said, because officials "kept coming to my house and asking me why I wasn't working. Eventually I came to look at Nhi Xuan and it seemed relatively all right."

The Assault Youth who tackle the back-breaking work of clearing land are recruited, for the most part, from the urban unemployed.[147] Most are school drop-outs. At least some have criminal backgrounds. One success story cited by Assault Youth leaders was Vu Hoang Vi. He was a thief, a heroin addict and a strong-arm thug, was drafted into the old Saigon army and sent to Quang Tri, the northernmost city controlled by Saigon. When Saigon lost Quang Tri in 1975, Vi returned to Saigon and continued his life as a thief. But after revolutionary forces took Saigon, theft became more difficult because the city was better organized. Despairing, Vi tried to hang himself, but that hurt too much. So he was persuaded to join the

Assault Youth. As part of the group's regimen, he studied before work every day, finished the ninth grade, and became deputy leader of his work group.

Whether or not this story is true, it does illustrate the ideals of the Assault Youth leaders. Members of the group serve for three years, and accept what amounts to military discipline.

Besides the NEZs in the Ho Chi Minh City area and along the Kampuchean border, zones are being established and expanded in Minh Hai province deep in the Mekong Delta and in the highland areas of northern and central Vietnam. The highlands NEZs will concentrate on cash crops, though they are also supposed to help make their provinces self-sufficient in food. In the central province of Dac Lac, more than 140,000 new settlers had arrived from various provinces along the central coast and from Thai Binh in the Red River Delta by the mid-1982.[148] They settled in seven NEZs, which concentrated on food production, eight state farms specializing in rubber and coffee, two logging camps and six farming co-ops.

Current plans call for a million more people to be moved to new economic zones between 1983 and 1985.[149] This may prove to be an overly-optimistic goal. But it is clear that "redeploying" the country's population will remain on important and necessary part of Vietnam's economic planning for some years to come.

RECONSIDERING THE FIRST FIVE YEAR PLAN

Reforms

When Vietnam's Political Bureau met in the summer of 1979 to look for ways to stimulate the country's lagging production, they adopted a startlingly direct approach. Those who produce more will earn more. It was an approach certain to raise protests from both doctrinaire socialists and egalitarian idealists. To lay the ideological groundwork, the party daily *Nhan Dan* argued that "only corpses do not need material benefits."[150] It attacked "fallacies" which contend that "mentioning material benefits is to speak of capitalism, that revolutionaries are advised to speak only of spiritual motives, and that revolutionary will can decide everything. . . ."

"Communists," *Nhan Dan* continued, "though treasuring spiritual motives, have always directed mass action toward the historical tide in keeping with objective laws. Worshipping spiritual motives and pursuing voluntarism, a product of idealism, is a reactionary ideology very harmful to the revolutionary cause of the masses." The scientific course, the materialist course, in other words, is to stimulate the "revolutionary enthusiasm of the masses" with some down-to-earth benefits.

Although the details are a subject of continuing discussion, the outlines of the policies adopted by the Political Bureau and later ratified by the 1982 party congress, involved:

- Contracts in agriculture. Work groups, usually families, agree to produce a certain amount from a section of land. The government, in turn, agrees to supply certain inputs such as seed, fertilizer and pesticides. Any surplus produced above the contracted amount belongs to the work group, which may keep it or sell it, either on the free market or to the state. This basic arrangement applies to cooperatives. Similar incentive systems were developed for state farms. Individual farmers, mainly in the southern Mekong Delta, were offered higher prices for their produce and could receive scarce agricultural inputs or consumer goods in barter arrangements.
- Piece-work rates in industry. Besides instituting pay based on productivity, the new measures allow increased autonomy to factory managers. If the state can not supply raw materials from non-government sources, or they can find no source of raw materials for the products they were scheduled to produce, they are allowed to shift to a line of products for which they can find raw materials.

Among the other measures were modest increases in pay for those on fixed salaries—primarily government cadre—and permission for some localities to set up their own import-export companies. There was also strong encouragement of small industries and handicraft enterprises to produce consumer goods. These small shops were seen partly as a way of soaking up urban unemployment but more importantly as a source of consumer goods needed for agricultural trade.

A major intended market for the consumer goods was the Mekong Delta, potentially the country's main rice basket. Farmers there had little incentive to grow more than enough for their subsistence needs, since there was little they could buy if they had a higher income. This meant that many of them did not see any particular advantage to double or triple cropping or to using the more troublesome high-yield strains. In another concession to farmers in the southern Delta, the goal of complete collectivization by 1980 was cancelled by the Political Bureau in 1979. The 1982 party congress emphasized a more gradual approach to collectivization. The emphasis at first would be on "transitional forms" such as labor-exchange groups (often formed only for one specific task) or marketing cooperatives. The target date for this scaled-down collectivization drive was postponed until 1985.[151]

Initial results of the new approach were promising. Production of staple crops rose steadily in 1981 and 1982. The 1982 harvest, more than 16 million tons, set a new record. It was 13 percent above 1980 production, and even exceeded the plan targets. Cash crops and livestock also showed healthy increases. Industrial output rose at a rate of 12.7 percent per year. (The annual rate in the 1976-1980 period averaged only 0.6 percent.)

Debates

This did not mean the reforms were set in motion without resistance. One member of the political bureau was said to have urged his home province not to implement the measures.[152] Another was reported to have said the only way to solve the food problem was to confiscate all the rice and distribute it equally. The continuing sharp debates were probably responsible for the long delay in convening the party's Fifth Congress, which finally met in March 1982, more than a year into the 1981-1985 plan period. The debates may even have prevented formal adoption of a five-year plan. No plan for the current period has been adopted, only the "directions, tasks and main objectives" have been outlined.[153]

Proponents of a less free-wheeling approach have been able to point to "abuses" of the new regulations such as speculative trading by marketing co-ops. Reports in *Nhan Dan* seldom offer much detail, but the type of transaction under criticism takes advantage of the fact that the co-op is authorized to buy raw materials directly and sell products directly, without going through government agencies. When the co-op instead buys a commodity, then resells it for a profit, it uses measures designed to stimulate production, but produces only inflation.

A related case reported by a western observer involved the Ho Chi Minh City import-export company, which was doing a quarter of its business in luxury goods sold for foreign currency to residents of the city.[154] "The picture of a state company perilously close to wholesaling to the black market has led to a recent government decision to control the companies more closely," the report said.

One point clarified at the 1982 party congress was that, while private initiative was welcomed in production, the government wanted to bring all trade under control of the state sector. The purpose is to ensure adequate supplies of essential goods at reasonable prices. But as late as the summer of 1983, Vietnamese officials admitted that the free market still controlled about 70 percent of the goods in circulation—the same proportion it did five years earlier.[155]

Collectivization in the Mekong Delta had faced similar resistance. Even though "transitional" forms are now included in the totals, government figures for late 1982 counted only 21.3 percent of farming families working 15.6 percent of the land as members of some form of collective.[156] This is about the same as figures commonly quoted in 1980, though the official statistics now indicate that only nine percent of the families were in collectives as recently as late 1981. The number of true cooperatives is still quite small—less than 200 in early 1983.

Cooperatives show higher yields—an average of 1.4 tons per hectare greater than individual farmers—but it is not clear how much of the difference can be attributed to the government's preferential treatment of co-ops. Cooperatives have a higher priority than individual farmers in the distribution of government-supplied fertilizer, pesticides and fuel and advice from government agricultural specialists. In any case, the government has a second purpose in encouraging co-ops. It is easier to keep track of their production and bring it into the state distribution network.

The trick Vietnam's leaders are trying to perform involves turning individual energies loose for producton and simultaneously reining in such "negative phenomena" as corruption and speculation. For several months in 1983 , the party daily carried the standing headline "Responsibility and Discipline." The front page articles stressed themes like meeting production targets, selling contracted amounts of food (or more than the contracted amounts) to the state, and paying taxes. On the other side of the balance, there were articles stressing the importance of the "family economy."

One such article, in the July 1983 issue of the party's theoretical journal, was written by "Truong Son," almost certainly a pseudonym for a high party official.[157] "A socialist economy," the author wrote, "includes three sectors: the state enterprise economy, the collective economy and the family economy." Bolstering his argument with references to Hungary and the Soviet Union (". . .the family economy in the Soviet Union accounts for 25-27 percent of agricultural production. . ."), Truong Son stressed the family economy's production of vegetables, fruit, meat, eggs and milk, both for family consumption and for sale. In rural areas, even on state farms families have their own private plots, which they are encouraged to use for small orchards, vegetable gardens, raising pigs and chickens, even for fish ponds. In the cities, the family economy includes handicraft production—knitting, sewing and embroidery, for example—and subcontracting to produce parts for light industrial co-ops.

An article in the previous month's edition of the journal argued against attempts to impose too much central planning on the economy.[158] It called for a balance between conflicting tendencies to allow market forces to control or, on the other hand, to attempt to preserve the "old order" of bureaucratic management. Because Vietnam's economy is so complex, involving interactions among so many small-scale production units, it is impossible to manage it centrally, the author said. The important thing was to delegate decision-making authority to an appropriate level, because when too many higher-level managers are second-guessing a manager at the working level, everyone may disclaim responsibility.

The tone of these articles suggests that Vietnam's pragmatists are not ready to yield to the counter-attack by hard-liners. The flexible policies rescued the country from the postwar crisis. The pragmatists will argue that, with some corrections to guard against abuse, these policies can be continued.

Even if production continues to rise as promisingly as it did in 1981 and 1982, there will be little immediate improvement in people's lives. Given the heavy burden of debt which Vietnam has already accumulated, much of any increased production must go to reduce imports and expand exports. This is already happening. When production of rice and other staples rose, grain imports were practically eliminated. But the country's diet remains below the poverty-line ration of 14 kilograms of rice a month. A survey of one farming cooperative in the northern Delta found the average adult had 10 kilograms of staples a month, plus vegetables, 200 grams (less than a half pound) of meat and one egg.[159]

In such difficult times, hope must be taken from small signs. Surely it is possible to draw real encouragement from the fact that admission requirements for one of Veitnam's more exclusive "clubs" have been tightened. In the war years, the North publicized the "five ton club" to recognize units which harvested a total of five tons of paddy per hectare from its two rice crops each year. Today, units may have to plant three crops a year to meet admission standards. Now it is the "ten ton club." The membership of that club, if it continues to grow fast enough, may make it possible to raise the diet from its current national average of 1,800 calories to the goals of 2,200 by 1990 and 2,500 by 2000.[160]

AMERICAN AND SOVIET RESPONSES

In 1978, when China canceled its aid projects in Vietnam, Hanoi was left with no immediate plan to replace the capital's congested, war-damaged, rail and highway bridge across the Red River. Soviet engineers stepped in to redesign the bridge, which was to have been one of China's major contributions to postwar Vietnam, and Soviet aid is paying for construction of the five-kilometer span. In 1981, when European companies exploring for oil off Vietnam's southern coast failed to reach agreement with the Vietnamese government on terms for a second phase of explorations, the Soviet Union was waiting. A joint Soviet-Vietnamese enterprise based in the coastal resort of Vung Tau hopes to complete its geological assessment of two exploration blocks by 1985.

Soviet resources are limited, of course, and not able to support every project which might have been planned with Chinese or western aid. But it has been quick enough and visible enough in picking up discarded projects and meeting crisis needs—like sending $2 million in relief aid after a 1982 typhoon—to accumulate an impressive amount of political capital.[161] It would be tempting to conclude that the main beneficiary of western and Chinese efforts to put the economic squeeze on Vietnam has been the Soviet Union.

United States Continues Resistance

A real thaw between the United States and Vietnam had begun in the first years of the Carter administration, but the atmosphere was frozen again, at least partly because American officials decided they would have to choose between China and Vietnam, that they could not normalize relations with both at once.[162] China won. About the same time, in December 1978, Vietnam sent its troops into neighboring Kampuchea to topple the government of Pol Pot. Attacks by Kampuchean troops on Vietnamese farming villages in the border region during 1977 and 1978 had seriously threatened Vietnamese efforts to reopen farmland deserted during the war. The United States, like China, has used the presence of Vietnamese troops in Kampuchea as its major talking point in efforts to persuade its allies and international organizations not to provide assistance to Vietnam.

In mid-1981, when American officials were lobbying vigorously to cut United Nations humanitarian and development assistance to Vietnam, the *Christian Science Monitor* reported that the United States was trying to "punish Vietnam for its illegal occupation of Cambodia."[163] The phrasing in a letter from the Commerce Department's Archie Andrews to the Mennonite Central Committee, also in 1981, was more diplomatic. He spoke of "the United States government's view that the Vietnamese government has within its power the ability to alleviate the severe hardship its policies have brought about."[164] At the time, the administration was blocking a Mennonite shipment of 250 tons of wheat flour to help meet Vietnam's pressing food needs. The shipment was eventually allowed, but the delay made the Reagan administration's point.

At the United Nations, Washington was particularly concerned about U.N. Development Program plans to help restore and improve Vietnam's overburdened transportation system.[165] A major worry was the program to restore 26 locomotives for the southern branch

61

of the railroad and to train Vietnamese technical and engineering personnel in the process. The United States was arguing that with a stronger southern division, Vietnam would be able to move cargoes by rail "right up to the Cambodian border." So far, the feeder line from Ho Chi Minh City to Loc Ninh in the rubber-producing border region has not been restored, and there is no other portion of the rail system that runs in the direction of Kampuchea. Nor does the Kampuchean railroad have any lines that approach the Vietnamese border. The UNDP staff noted that "railways constitute the only dependable means" of transport in a long country with northern and southern regions which are "complementary and dependent on each other" for national development.

Despite the American lobbying, the UNDP approved the railroad program as part of a package of 54 projects costing $94 million. But the United States succeeded in having the projects delayed a year for review—again sending a sharp political message to Hanoi, and in this case at least temporarily blocking assistance which represented a significant boost for the Vietnamese economy.

It is difficult to assess the American contention that Vietnam's economy would benefit from a withdrawal of its forces from Kampuchea. Direct costs to Vietnam of keeping its troops in Kampuchea have been estimated at $11.7 million a year (excluding the cost of weapons and munitions, which the Vietnamese say are supplied gratis by the Soviets)—about 0.6 percent of the national budget of some $1.9 billion in 1981.[166] Vietnam has lost the services of some people with managerial skills, now officers in Kampuchea. But the ordinary soldiers might be more of a burden than a blessing if released in a Vietnamese economy still suffering significant unemployment.

The American campaign has had mixed effects. Japan and Australia suspended aid, though their trade with Vietnam has continued. The Asian Development Bank has not approved any new projects since 1978. In addition to the tiny trickle of aid from a few private U.S. agencies such as Oxfam America and AFSC, the Scandanavian countries have sent aid, as have Belgium, France, West Germany and the United Kingdon. But by far the largest source of reconstruction and development assistance has been Moscow.

Soviet Aid

At the end of the war, the Soviet Union cancelled Vietnamese debts of $450 million.[167] American estimates of current Soviet aid to Vietnam generally use figures from $3 million to $6 million a day — or $1 billion to $2 billion per year. A recent NATO estimate of Soviet aid in 1980 suggests a figure somewhat less than $1 billion, and notes that by far the dominant part of the aid was economic rather than military.[168] A British assessment puts economic aid at some $840 million in 1980, $940 million in 1981.[169]

Much of the Soviet assistance has gone toward building up Vietnam's power generating capacity. The most ambitious project is a combined water control and hydroelectric dam on the Black River [Da River], the major tributary of the Red River, near Hoa Binh southwest of Hanoi. When it is completed, it will have a capacity of nearly two million kilowatts. Surveys for the project began in 1970 and accelerated after the end of the war. In 1981, construction of a diversion channel for the river, more than a kilometer long, began, and in January 1983 that channel was complete. The flow of the river was shifted to its temporary course, and huge trucks began moving the cement blocks, rocks and earth to close a dam which will be 128 meters high and more than 800 meters long. The first generator is scheduled to go into operation in 1987.[170]

Another major hydropower project is planned for the South, near Bien Hoa and a thermal power plant is under construction at Pha Lai in the North.[171] Other major projects in the list of 40 already underway include a diesel motor plant, a mining equipment repair facility, two new coal mines, a factory casting concrete panels for prefabricated housing and a cement plant. The Soviet Union has also invested in expanding Vietnam's rubber, coffee and tea production.

One of the more controversial aspects of Soviet aid has been its program to train technicians and skilled workers. The U.S. State Department, apparently on the basis of reports by anti-Hanoi Vietnamese emigres, has charged that between 100,000 and 500,000 Vietnamese workers would be sent to the Soviet Union in the 1981-1985 period and that there had been reports of "harsh — and, in some cases, involuntary — conditions" for the workers.[172] Some sensationalized versions of the story allege half a million Vietnamese prisoners in Siberia working on the Soviet oil pipeline.

Vietnamese and Soviet responses have stressed the long-term program to train Vietnamese in technical skills, which includes Soviet and Eastern European instructors in Vietnam as well as Vietnamese in Eastern Bloc schools and factories. Hanoi sources put the number of workers in all bloc countries in 1982 at 50,000 and said the number was expected to double in the next four years.[173] Western diplomats in Hanoi say that the Soviet and East European work assignments are so coveted that young workers have to offer bribes to get them.[174] The attraction is higher wages and the opportunity to buy consumer goods unavailable in Vietnam as well as skills which will qualify these workers for relatively high-paying jobs in Vietnam when they return.

The main thing the Soviets appear to get in return for their aid to Vietnam is an ally in a strategically important part of the world. The Vietnamese workers are probably a help to the Soviet Union, which seems always to be short of labor. Vietnam supplies fresh fruits and vegetables to the Soviet Far East. Some bloc countries are taking advantage of the relatively inexpensive labor in Vietnam's textile mills. Some Vietnamese rubber goes to the Soviet Union, as do some light industrial and handicraft products. But the overall trade balance is hundreds of millions of dollars out of balance each year. And in any case, Soviet loans covering the deficits carry interest of no more than two percent.[175] What must be more important to the Soviets is their access to such facilities as the U.S.-built naval base at Cam Ranh and air field in Danang. American sources say the Soviets have even built an electronic intelligence complex at Cam Ranh, which permits them to monitor communications to American installations in the Philippines.[176]

Soviet aid is not limitless. A Soviet official recently stressed that Vietnam needs to "turn to account in a more rational way the production potential already built...."[177] And observers have noted a change in Soviet aid—giving assistance on a project-by-project basis rather than providing an overall package for the whole five-year plan period.[178] Still, Moscow remains Hanoi's only steady ally. Vietnamese statements indicate that they return the loyalty. This must raise questions about whether American policy is not weakening a small adversary in the region only to strengthen a larger one.

1. Committee on the Judiciary, United States Senate, *Relief and Rehabilitation of War Victims in Indochina: One Year After the Ceasefire*, GPO, Washington, January 27, 1974, p. 6.

2. International Bank for Reconstruction and Development, *Current Economic Position and Prospects of the Republic of Viet Nam*, January 18, 1974, p. 2. (Hereinafter referred to as IBRD.)

3. Ibid., p. 24.

4. John Spragens, Jr., interview with Le Quang Chanh, Deputy Chairman of the Ho Chi Minh City People's Committe, August 11, 1980.

5. William J. Lederer, *Our Own Worst Enemy* (New York: Norton, 1968), p. 97.

6. IBRD, p. 2.

7. Ibid., p. 3.

8. Patrick M. Boarman, "The Economic Promise of South Vietnam: An

Overview," *The Economy of South Vietnam: A New Beginning*, (Los Angeles: Center for International Business, 1973), p. 14.

9. "The Sputtering Honda Economy," *Newsweek*, October 15, 1973, p. 48.

10. IBRD, p. 14.

11. Ibid., p. i.

12. Boarman, p. 17.

13. Ibid., p. 31.

14. Ibid., p. 12.

15. Bryan Frith, "Vietnam: Explorer Gets a Toehold," *Far Eastern Economic Review*, November 21, 1975, p. 63.

16. John Spragens, Jr., "Corruption in South Viet Nam: Pervasive, Profound and Permanent," *American Report*, July 22, 1974.

17. IBRD, table 6.

18. John Spragens, Jr., "'Desperate' Food Crisis in Danang," *American, Report*, April 29, 1974.

19. John Spragens, Jr., "Down in the Valley: Two Weeks with the PRG," *American Report*, July 8, 1974.

20. Alexandre Casella, "The PRG's Vietnam," *Le Monde*, December 19 and 20, 1974.

21. Ibid.

22. Le Hong Tam, "35 nam xay dung va phat trien nen cong nghiep xa hoi chu nghia" [35 Years of Building and Developing a Socialist Industry], in *35 Nam Kinh te Viet Nam (1945-1980)* [35 Years of the Vietnamese Economy], Hanoi, Nha Xuat ban Khoa hoc Xa hoi, 1980, p. 78.

23. Ibid., p. 94.

24. Nguyen Huy, "35 nam thuc hien duong loi phat trien nong nghiep cua Dang" [35 years of Implementing the Party's Line on Agricultural Development], in *35 Nam Kinh te Viet Nam*, p. 155.

25. "Results of North Viet Nam's Second Census," *Vietnam Courier*, Hanoi, October 1974, p. 7.

26. Pham Van Dong, "19 Months of Economic Rehabilitation and Development in North Viet Nam," *Vietnam Courier*, Hanoi, October 1974, p. 3.

27. General William C. Westmoreland, *A Soldier Reports* (New York: Doubleday, 1976) p. 340.

28. Ibid., p. 341.

29. David Burnham, "1965 Memos Show Dow's Anxiety on Dioxin," *New York Times*, April 19, 1983.

30. Ralph Blumenthal, "Files Show Dioxin Makers Knew of Hazards," *New York Times*, July 6, 1983.

31. Burnham.

32. Ibid.

33. John Lewallen, *Ecology of Devastation: Indochina* (Baltimore: Penguin, 1971), p. 115.

34. Ibid., p. 116.

35. Ibid., p. 117.

36. "Latest Research on the Problem of Mutagenic Effects on the First Generation after Exposure to Herbicides," *US Chemical Warfare and its Consequences* (Hanoi: Vietnam Courier, 1980), p. 73ff.

37. Dan Thu, "Dan so Viet nam" |Vietnamese Population|, *Doan Ket*, Paris, April 1983.

38. Nguyen Khac Vien, "The Lasting Consequences of Chemical Warfare," *Herbicides and Defoliants in War: The Long-Term Effects on Man and Nature*, (Hanoi: Vietnam Courier, 1983).

39. Ibid.

40. "Agent of Mutilation?" *Asia Week*, March 14, 1980.

41. John Lewallen, p. 115 and Vien. Information in this section is based on these two works except as otherwise noted.

42. Vu Tu Lap, *Vietnam: Geographical Data*, (Hanoi: Foreign Language Publishing House, 1979), p. 95ff describes this and other ecosystems.

43. John Spragens, Jr., interview with Nguyen Khac Vien, August 5, 1980.

44. The text of the "Agreement on Ending the War and Restoring Peace in Viet Nam" can be found in the *New York Times*, January 25, 1975.

45. "Documents Concerning the US Government's Pledge to Contribute to Healing the Wounds of War and to Post-war Reconstruction in Viet Nam," *Vietnam Courier*, Hanoi, June 1977, p. 2.

46. Senator George McGovern, *Vietnam: 1976*, Committee on Foreign Relations, March 1976, p. 14.

47. Ibid. and Heidi Kuglin, "Vietnam Rebuilds—No Thanks to America," *Los Angeles Times*, August 16, 1976.

48. Quoted in McGovern, pp. 14-15.

49. IBRD, p. i.

50. McGovern, p. 16.

51. *Congressional Record—Senate*, May 14, 1975, p. S8152.

52. House Subcommittee on International Trade and Commerce, *United States Embargo of Trade with South Vietnam and Cambodia*, June 4, 1975, p. 3.

53. Ibid.

54. Ibid., p. 4.

55. Richard Borsuk, "Capitalizing on Communism: Businessmen Urge Viet Trade," Pacific News Service, December 18, 1975.

56. Ibid.

57. *Congressional Record—House*, October 1, 1975, p. H9433.

58. House Subcommittee on International Trade and Commerce, p. 1.

59. Nayan Chanda, "Speeding towards reunification," *Far Eastern Economic Review*, December 5, 1975.

60. Paul and Sophie Quinn-Judge, "Saigon, 'A Big Nut to Crack'," *Indochina Chronicle*, No. 44, October-November 1975.

61. Wilfred Burchett, "Vietnam to reunify very soon," (New York) *Guardian*, November 19, 1975.

62. Kathleen Teltsch, "U.N. Assembly, 123-0, Asks Reversal of Veto of Vietnams," *New York Times*, September 20, 1975.

63. *The Reunification of Viet Nam*, (Hanoi: Foreign Languages Publishing House, 1975), pp. 15-16.

64. *Viet Nam: Forward to a New Stage*, (Hanoi: Foreign Languages Publishing House, 1977), pp. 33-34.

65. Nayan Chanda.

66. General Van Tien Dung, *Our Great Spring Victory*, (New York: Monthly Review, 1977), p. 24ff.

67. John Spragens, Jr., interview with Nguyen Khac Vien, August 5, 1980.

68. "Urban Affairs Discussion, November 4, 1970," in *Indochina Chronicle*, No. 4, September 1, 1971, p. 4.

69. Seth Lipsky, "Vietnam Reds Struggle With Economic Woes, Corruption, Resistance," *Wall Street Journal*, December 22, 1975.

70. Earl S. Martin, *Reaching the Other Side*, (New York: Crown, 1978), p. 128ff.

71. Paul and Sophie Quinn-Judge, "Saigon, 'A Big Nut to Crack'."

72. "Saigon Tries Persuasion in Restoring Rural Life," *New York Times*, May 23, 1975.

73. Nayan Chanda, "Vietnam: Back to the Land," *Far Eastern Economic Review*, June 27, 1975.

74. Nayan Chanda, "Requiem for an Old Order," *Far Eastern Economic Review*, June 6, 1975.

75. "The Economic Situation in Ho Chi Minh City Two Years After Liberation," in *With Firm Steps*, (Hanoi: Foreign Languages Publishing House, 1978), p. 63.

76. Jean de la Gueriviere, "'Reactionary' Round-up in Saigon," *Le Monde*, July 17 and 18, 1975, in (Manchester) *Guardian* Weekly, August 2, 1975.

77. Nayan Chanda, "Cautious step towards revival," *Far Eastern Economic Review*, August 1, 1975.

78. Philippe Pons, "Disenchantment in Saigon," *Le Monde*, August 20, 1975, in (Manchester) *Guardian* Weekly, August 30, 1975.

79. Nayan Chanda, "Vietnam: Back to the Land."

80. Nayan Chanda, "Vietnam: Cash Withdrawal," *Far Eastern Economic Review*, October 3, 1975.

81. Wilfred Burchett, "Saigon Slams Profiteers," (New York) *Guardian*, September 24, 1975.

82. David A. Andelman, "Saigon Unit Seeks Diverted U.S. Aid," *New York Times*, September 28, 1975.

83. Ho Kwon Ping, "Lee: The reluctant middle-man," *Far Eastern Ecomomic Review*, October 17, 1975.

84. "Japan-N. Vietnam Aid Agreement Initialed," *Japan Times Weekly*,

October 11, 1975.

85. George McArthur, "Hanoi Tries to Lure American Oilmen," *Los Angeles Times*, December 14, 1975.

86. "Vietnam Looks to U.S. Attitude," *New York Times*, November 23, 1975.

87. "Ford Suggests Links to Hanoi Are Possible," *Los Angeles Times*, December 8, 1975.

88. "Saigon: Promoting the spirit of friendship," *Far Eastern Economic Review*, August 8, 1975.

89. "Saigon Completes Part of Rail Line Extending to Hanoi," *New York Times*, November 30, 1975.

90. Martin Woollacott, "Vietnam: still two nations," (Manchester) *Guardian* weekly, April 25, 1976.

91. "Main Targets of the Second Five-Year Plan," *Vietnam Courier*, Hanoi, January 1977.

92. Vo Thuc Dong, "21 Million Tons of Cereals by 1980: An Attainable Target," *Vietnam Courier*, July 1977, and *Statistical Data of the Socialist Republic of Vietnam 1978* (Hanoi: General Statistics Office, 1979). (Hereinafter referred to as *Statistical Data*).

93. "Outline of the Draft Political Report of the Central Committee of the Vietnam Workers Party to the Fourth Party Congress," *Vietnam Courier*, December 1976.

94. *Statistical Data*.

95. International Monetary Fund, *Socialist Republic of Vietnam - Recent Economic Development*, May 18, 1982. (Hereinafter referred to as IMF).

96. Nayan Chanda, "Bridging Hanoi's food gap," *Far Eastern Economic Review*, August 20, 1976.

97. "Moscow Expands Aid to Vietnamese," *New York Times*, February 1, 1976.

98. Nayan Chanda, "Vietnam's joint venture plan," *Far Eastern Economic Review*, September 24, 1976.

99. Malcolm Salmon, "Hanoi: After revolution, evolution," *Far Eastern Economic Review*, December 12, 1975.

100. Nayan Chanda, "Vietnam's joint venture plan."

101. "Regulations on Foreign Investment in the Socialist Republic of Vietnam," *Vietnam Courier*, Hanoi, July 1977.

102. *Indochina Spotlight* (Bank of America Asia Representative Office), July 1977.

103. Susumu Awanohara, "The Rush for Vietnam's Oil," *Far Eastern Economic Review*, February 20, 1976.

104. *Indochina Spotlight*, September 1977.

105. Barry Wain, "Vietnamese Oil Search: Exit Westerners, Enter Soviets," *Asian Wall Street Journal*, January 16, 1981.

106. Nayan Chanda, "Vietnam's battle of the home front," *Far Eastern Economic Review*, November 2, 1979.

107. Michael Morrow, "Vietnam's Embargoed Economy: In the U.S. Interest?" *Indochina Issues*, August 1979.

108. Bill Herod, "The Unfinished Business of America's MIAs," *Indochina Issues*, June 1981.

109. *Indochina Spotlight*, March 1977.

110. *Indochina Spotlight*, April 1977.

111. Gareth Porter, "The 'China Card' and US Indochina Policy," *Indochina Issues*, November 1980.

112. *Statistical Data*.

113. Henry Kamm, "Vietnam Issues a Gloomy Report on Economy in 1979," *New York Times*, December 27, 1979.

114. John Spragens, Jr., "Looking Ahead," *Southeast Asia Chronicle* No. 76, Decmeber 1980.

115. Nayan Chanda, "Hanoi takes the campaign behind the lines," *Far Eastern Economic Review*, March 3, 1978.

116. "Vietnam reported planning Cambodia knockout blow," *Dallas Times-Herald*, October 5, 1978.

117. Michael Morrow, "The move towards consolidation," *Far Eastern Economic Review*, June 20, 1975.

118. Van Son and Hoang Hien, "Quy mo hop tac xa o Ha Nam Ninh" [Scale of cooperatives in Ha Nam Ninh], *Nhan Dan*, April 7, 1980, and John Spragens, Jr., interviews in Vietnam, August 1980.

119. Tam Huyen, article from *Nhan Dan* broadcast by Hanoi domestic service, June 19, 1980, in Foreign Broadcast Information Service, June 19, 1980.

120. John Spragens, Jr., "Vietnam, China: Allies No More," *St. Louis Post-Dispatch*, September 21, 1980.

121. Nayan Chanda, "Vietnam's battle of the home front."

122. Nayan Chanda, "Comrades curb the capitalists," *Far Eastern Economic Review*, April 14, 1978.

123. Murray Hiebert, "Vietnam's Ethnic Chinese," *Southeast Asia Chronicle* No. 68, December 1979.

124. Seymour M. Hersh, "Exodus of Skilled Ethnic Chinese Worsens Hanoi's Plight," *New York Times*, August 9, 1979.

125. IMF.

126. *Nhan Dan*, October 10, 1979, in Foreign Broadcast Information Service, October 16, 1979.

127. "Japan will delay aid to Vietnam," *Philadelphia Inquirer*, March 13, 1980.

128. Ted Morello, "The US losses a Vietnam battle," *Far Eastern Economic Review*, July 10, 1981.

129. *Asian Wall Street Journal*, November 3, 1979.

130. Nayan Chanda, "Comrades curb the capitalist."

131. Murray Hiebert, "Vietnam's Ethnic Chinese."

132. Refugees: Still an Issue," *Southeast Asia Chronicle* No. 76, December 1980.

133. Murray Hiebert, "Vietnam's Ethnic Chinese."

134. "Refugees: Still an Issue."

135. Seymour Hersh, "Exodus of Skilled Ethnic Chinese Worsen's Hanoi's Plight," *New York Times*, August 9, 1979.

136. Murray Hiebert, "Vietnam's Ethnic Chinese."

137. Julie Forsythe and Tom Hoskins, *Visit to Pulau Bidong Refugee Camp*, American Friends Service Committee, May 1979.

138. John Spragens, Jr., interview in United States, March 1981.

139. *Report of the United Nations Mission to North and South Viet-Nam*, March 1976.

140. "Xay dung Thanh pho Ho Chi Minh City thanh mot thanh pho xa hoi cu nghia giau dep" [Build Ho Chi Minh City into a rich, beautiful socialist city], *Tap Chi Cong San* [Communist Review], November 1982.

141. *Information-Documents*, (Hanoi: Vietnam Courier), March 1, 1982.

142. John Spragens, Jr., "Looking Ahead," *Southeast Asia Chronicle* No. 76, December 1980.

143. "Chinh sach xay dung cac vung kinh te moi" [Policies on building new economic zones], *Nhan Dan*, April 12, 1980.

144. *Information-Documents*, (Hanoi: Vietnam Courier), March 1, 1982.

145. Paul Quinn-Judge, "Down on the state farm," *Far Eastern Economic Review*, June 1, 1983.

146. Ibid.

147. John Spragens, Jr., interview in Ho Chi Minh City, August 1980.

148. *Information-Documents*, July 1, 1982.

149. *Information-Documents*, January 1, 1983.

150. "Spiritual Motives, Material Benefits" (editorial), *Nhan Dan*, October 23, 1979, in Foreign Broadcast Information Service, October 26, 1979.

151. "Socio-Econoic Plan for 1983 and Efforts To Be Made Until 1985," Vietnam News Agency, December 22, 1982, in Foreign Broadcast Information Service, December 30, 1982.

152. Paul Quinn-Judge, "A touch of capitalism leaves Vietnam's communists in a quandary," *Christian Science Monitor*, March 9, 1983.

153. "Socio-Econoic Plan . . ." and Jacques de Barrin, "Market economy takes on the collectives," *Le Monde*, January 8-9-10, 1983, in (Manchester) *Guardian* weekly, January 30, 1983.

154. Paul Quinn-Judge, "A touch of capitalism . . . ."

155. Paul Quinn-Judge, "Ideological backtracking," *Far Eastern Economic Review*, July 21, 1983.

156. *Information-Documents*, (Hanoi: Vietnam Courier), February 16, 1983.

157. Truong Son, "Kinh te gia dinh" [The family economy], *Tap Chi Con San* [Communist Review], July 1983.

158. Che Viet Tan, "Doi moi cong tac ke hoach hoa va phan cap quan ly kinh te" [Change the work of planning and the division of responsibility in economic management], *Tap Chi Cong San*, June 1983.

159. Tran Van Ha, "Kinh te gia dinh va he thong V.A.C. o Viet Nam" [The family economy and the system of gardens, fishponds and animal raising in Vietnam], *Nhan Dan*, November 2-3, 1982.

160. Ibid.

161. *Information-Documents*, (Hanoi: Vietnam Courier), May 16, 1982.

162. Gareth Porter, "The 'China Card' and US Indochina Policy," *Indochina Issues*, November 1980.

163. Louis Wiznitzer, "US tries to punish Vietnam by paring UN assistance," *Christian Science Monitor*, May 26, 1981.

164. "Food Gift to Vietnam From Church Halted," *Washington Post*, May 28, 1981.

165. Ted Morello, "The US loses a Vietnam battle," *Far Eastern Economic Review*, July 10, 1981, and Ted Morello, "Railroading Vietnam," *Far Eastern Economic Review*, May 21, 1982.

166. Nayan Chanda, "Vietnam's Economy: 'Bad, But Not Worse," *Indochina Issues*, October 1983.

167. IMF.

168. David Fouquet, "Aid to Cuba, other clients may have hit limit," *Christian Science Monitor*, January 4, 1983, and "Third World Aid Called Drain on Soviet Resources," *New York Times*, January 23, 1983.

169. "Vietnam: Economy in Difficulties, Labor Exported," Foreign and Commonwealth Office, London, October 1982.

170. "Blocking the Stream of the Da River," *Vietnam Courier*, Hanoi, February 1983.

171. *Information-Documents*, May 16, 1982.

172. William Branigin, "Vietnamese, Contrary to U.S Charges, Said to Covet Soviet Bloc Jobs," *Washington Post*, March 13, 1983.

173. Nayan Chanda, "Now, the 'flot' people," *Far Eastern Economic Review*, May 14, 1982.

174. William Branigin, "Vietnamese, Contrary to U.S Charges, ..."

175. IMF.

176. "Soviet Is Making More Use of Cam Ranh Bay," *New York Times*, March 13, 1983.

177. M.S. Gorbachev, speech to Fifth Congress of the Vietnamese Communist Party, March 28, 1982, in Foreign Broadcast Information Service, March 31, 1982.

178. Paul Quinn-Judge, "Rewards of modesty," *Far Eastern Economic Review*, February 10, 1983.

CHAPTER THREE
Kampuchea

Since 1970 the predominantly rural and isolated peasant society of Kampuchea* has undergone a period of shocking stress, brutality, and change. The national experience of Kampuchea since that date includes multiple and massive invasion by foreign troops, prolonged civil war, large-scale aerial bombardment, mass enslavement of the population in the name of Khmer nationalism and self-reliance, occupation by a traditional enemy, and famine. In the 1970s between 1.5 and 2 million Khmer out of a population of approximately 7.3 million lost their lives as a consequence of war, revolution, mass murder, and starvation. The food system of Kampuchea, which had endured essentially unchanged for hundreds of years, was shattered. Today, while a guerrilla war still continues, the peasants of Kampuchea have begun to piece together this system under new and difficult circumstances.

---

*The authors use "Kampuchea" to designate the country known to Americans as "Cambodia." Kampuchea is a different rendering of the actual name of the country in Khmer, the national language of Kampuchea. The Khmer are the country's major ethnic group, making up about 90 percent of the population. The myth of the origin of the Khmer people involves "Kambuja," the Sanskrit name of a tribe in northern India; "Kampuchea" derives from the name of this tribe.

In the context of recent history, however, the name Kampuchea has taken on political significance. After independence the official name of the country ruled by Prince Norodom Sihanouk was the Kingdom of Cambodia. After the Lon Nol coup in 1970, the country became the Khmer Republic. In 1975 the victorious Khmer Rouge changed the name to Democratic Kampuchea on the grounds that "Cambodia" was an inauthentic legacy of colonialism. After the Vietnamese invasion in late 1978, Heng Samrin assumed power of the People's Republic of Kampuchea. Thus "Kampuchea" is associated with recent revolutionary movements, including one (the Khmer Rouge) responsible for the death of at least one million Kampucheans.

The United States has contributed inordinately to this on-going tragedy. The United States bombed the border areas of Kampuchea secretly in 1969, without informing the American people. In 1970, the joint U.S.-South Vietnamese attack on North Vietnamese bases in Kampuchea drew that country inextricably into the Indochina conflict. The United States bombed the Kampuchean countryside in support of the Lon Nol government from 1970-73, a bombing campaign so massive and brutal that the peasants of Kampuchea have yet to recover from its effects.[1] Yet even during the Lon Nol period, the United States refused to supply adequate humanitarian aid to mitigate the consequences of the war effort. The amount of military assistance to the American client government led by Lon Nol dwarfed that of humanitarian aid.[2]

When the Lon Nol government fell to the Khmer Rouge in April 1975, the United States immediately imposed a trade embargo on Democratic Kampuchea which has been maintained to the present. Only in the aftermath of the Vietnamese invasion of Kampuchea, which toppled the Khmer Rouge in early 1979, has this embargo been relaxed enough to allow American humanitarian assistance to reach Khmer people in need. As the earlier threat of *famine* faded in 1981, however, U.S. policy came closer to the reimposition of a complete trade embargo against the new People's Republic of Kampuchea. The trade embargo, which blocks not merely trade but essential humanitarian assistance, threatens to bring a premature halt to the recovery of the Khmer people from the disaster of the 70s.

American private aid agencies have provided a humanitarian counterpoint to the destructive U.S. government involvement in Kampuchea's recent history. Even before 1975, during the civil war between Lon Nol and Pol Pot forces, agencies such as CARE, Catholic Relief Services, and World Vision attempted to provide primarily emergency food and medicine to the millions of displaced

---

For this reason, to some Kampucheans, particularly exiles, "Kampuchea" suggests revolution and death, "Cambodia" the halcyon days when their country was whole. Khmer refugees in the United States, however, sometimes refer to their homeland as "Kampuchea."

In this work Kampuchea is used as the name of the country throughout history, even for periods when it was formally known as Cambodia. This choice derives solely from the fact that the present rulers of the country call it Kampuchea. "Kampucheans" or "the Khmer people" are used interchangeably with "the people of Kampuchea." See the Preface of David P. Chandler's *A History of Cambodia* (Boulder, Colorado: Westview Press, 1983), for a discussion of this issue.

persons seeking refuge in towns from the brutal war in the country-side. But this crisis, as great as it was, was compounded by the rule of the Khmer Rouge from 1975 to early 1979. The Khmer Rouge attempted to solve the food crisis created by the civil war through the mass mobilization of the entire population of the country on agricultural reconstruction projects. But at the same time, the leadership sought to punish those in the urban *and* rural parts of the country who had failed to join the Khmer Rouge before the Pol Pot victory. These latecomers, the so-called "New People," were denied adequate food and medical care; they were therefore decimated by starvation amd disease brought on by the hard labor under slave conditions. The leader of the Khmer Rouge, Pol Pot, brought his country to the point of collapse by daring to challenge its more powerful neighbor, Vietnam, for control of the Mekong Delta region of southern Vietnam.

In 1979, after Vietnamese troops and a small contingent of Kampuchean rebels toppled Pol Pot, what remained of Khmer society hemorrhaged. The spectacle of skeletal survivors of the Khmer Rouge holocaust (including Khmer Rouge cadre) reaching refugee camps in Thailand, and reports of similar masses of people criss-crossing Kampuchea seeking lost relatives and returning to their home villages from work camps, gripped the world, and forced international aid agencies to respond to the crisis. While slow to get underway, the Kampuchea relief effort, once begun, was the largest outpouring of private humanitarian aid in the brief history of such undertakings in this century. American agencies working both in Kampuchea and at the Thai border raised millions of dollars from private donations in the United States for relief and rehabilitation programs. Oxfam America, which in its 1978 fiscal year had an operating budget of $472,000, raised more than *$4.2 million* in less than a year for its Kampuchea program. The commitment of American agencies such as Oxfam America to the long-term well-being of the people of Kampuchea dates from the intense period of involvement in the famine relief effort which began in August 1979.

The people of Kampuchea, like those of Biafra or Bangladesh, have suffered misfortune on such a scale that an international relief effort has become a part of its modern history. For Kampuchea, however, the length and intensity of its suffering and the political conflict which continues to engulf it mean that the disaster has had an impact which cannot be erased by a single aid effort, however massive. It will take at least a generation for the people of Kampuchea to recover from the human and physical destruction of the 1970s. The legacy of this disastrous decade is with the people of Kampuchea every day—in the memories of lost family members, in the lack of draft animals and other means of production, in the potholed roads and damaged machinery, in the thousands of children growing up without parents, in the lack of trained people at all levels required to make a modern country function.

After witnessing the remarkable recovery of the Khmer people from the worst of the famine period in 1979 and 1980, international relief workers and to some extent the people of Kampuchea themselves deluded themselves into thinking that full recovery would be swift. The revival of markets, the revival of Buddhist rituals, the many children being born as families reconstituted themselves became symbols of the sudden, phoenix-like rebirth of the Khmer people. And compared to the horrors of the Pol Pot period and the famine, the recovery *was* remarkable.

But since early 1981, the pace of recovery has in fact been painstaking. After a 1983 rainy season rice crop plagued by drought and flooding, the estimated 1984 rice deficit for Kampuchea, based on a *minimum* need of 12 kg per person per month, stood at almost 300,000 tons—virtually identical to the food deficit after the 1980 rainy season harvest, the harvest which ended the famine.[3] According to UNICEF, Kampuchea has the fourth highest infant mortality rate in the world, with 20 percent of its children dying before the age of one.[4] The 1983 World Bank figures list Kampuchea as the poorest country in the world. These statistics, among many others, suggest that the people of Kampuchea have barely reached a subsistence level in their struggle to overcome the legacy of the 1970s.

The difficult situation faced by the people of Kampuchea today gives urgency to this Impact Audit. At the very time when the long-term problems faced by the people of Kampuchea are becoming quite clear, the U.S. government has ended its contribution to aid programs inside the country, and is moving to block essential humanitarian aid programs of Oxfam America, other American voluntary organizations, and some international agencies as well. The people of Kampuchea have not yet recovered from the terrible man-made disasters of the 1970s. To cut them off now from outside aid, even aid donated by private American citizens, is cruel and unusual punishment indeed for people who have been victims once too often of U.S. government policy.

HISTORICAL BACKGROUND

Historically, the power of Kampuchean civilization reached its zenith in the 10th-13th centuries when an empire based on a sophisticated system of rice irrigation centered at the temple city of Angkor Wat completed the conquest of neighboring states in what is now southern Vietnam and Thailand.[5] The decline of this state due to poor leadership and the resulting breakdown in the maintenance of the irrigation system ushered in a period of powerlessness which has continued virtually unbroken to the present day. Burmese, Thai, and Vietnamese armies invaded Khmer territory, sacked the great temples, and imposed systems of tribute on the weak Khmer monarchy from the 15th to the late 18th centuries. In the 1840s the Vietnamese had virtually complete sovereignty over eastern Kampuchea, while the Thais controlled the western provinces. The divided monarchy was completely beholden to one or the other of the two powerful neighbors. The first half of the 19th century was comparable to the 1970s for its level of chaos and suffering.[6]

The French intervention in 1863 at the behest of Kampuchea's King Norodom salvaged only a measure of Kampuchean sovereignty. French colonialism may have forced Vietnamese and Thai troops out of the country, but it substituted a colonial administration staffed by Vietnamese civil servants under French tutelage.[7] The modernizers of Kampuchea, therefore, were not the Khmer themselves but French and Vietnamese administrators and the Chinese merchant class which dominated Kampuchea's commerce.[8] Thus, when Prince Norodom Sihanouk, then only 31 years old, siezed the moment from more radical politicians in the early 1950s and declared Kampuchea's independence, the country had still only a very tiny Kampuchean elite able to govern. Further, despite high levels of taxation on the peasantry, there had been virtually no investment in rural areas by the colonial administration.[9] France had extracted huge rice surpluses through taxation, but peasants lived at the traditional, meagre subsistence level. Industrialization was virtually nonexistent. In addition to rice exports, rubber plantations provided a sure source of foreign exchange, but one exploited primarily by private French firms such as Michelin, which repatriated their large profits rather than investing them in Kampuchea.[10]

Initially, Sihanouk played the traditional kingship role to perfection. While peasant allegiance to Sihanouk's personal, emotional leadership was particularly strong, this style of leadership, which derived from the semi-devine status traditionally conferred upon royalty, became increasingly anachronistic in the polarized Indochina context of the 1960s.[11] In the international arena Sihanouk developed a concept of neutrality which tried to keep Kampuchea clear of the conflict in Vietnam. He accepted aid from a broad range of sources: French, Czech, Yugoslav, Australian and West German, in addition to Chinese, Soviet and American. Yet Kampuchea's weakness ultimately led to the betrayal of this neutrality. Sihanouk supported the struggle of the North Vietnamese and the National Liberation Front (NLF); he tolerated the use of Kampuchean territory by their forces. Further, one important branch of the Ho Chi Minh Trail was not a jungle path at all but a sea route from China to Kampuchea's main southern ocean port, Sihanoukville (now Kompong Som).[12] From this port trucks carried arms and supplies about 100 miles overland to border enclaves. As the intensity of the war increased in the 1960s, the U.S. frustration with the use of Kampuchean territory by the North Vietnamese and NLF increased accordingly.

U.S. pressure on Sihanouk to allow military operations inside Kampuchea against the Vietnamese resistance coincided with domestic frustration with Sihanouk's approach to governance. Corruption was rampant in the latter part of his rule and the economy was near breakdown.[13] The left, the core of the group which formed the Khmer Rouge, tried initially to work within the Sihanouk system, but by 1967 the leadership had slipped into the jungle to begin working for revolutionary change.[14] Sihanouk's brutal suppression of a peasant revolt in Battambang in the same year demonstrated to the radicals that they could only survive in armed opposition. In the cities pressure on Sihanouk took the form of demonstrations by students and civil servants agitating to abolish the trappings of monarchy, all vestiges of feudal rule, and to create modern, rational democratic political institutions. The combination of domestic political pressure and the rapidly widening Indochina War (the secret bombing of Kampuchea by the United States began in March 1969) destroyed Sihanouk's fragile neutral consensus and produced the coup against him in March 1970 which placed the right wing pro-American General Lon Nol in power.

CIVIL WAR

Lon Nol immediately entered Kampuchea into the Indochina War on the side of the United States and its South Vietnamese allies. With Lon Nol's blessing, a combined U.S.-South Vietnamese force invaded Kampuchea in May to clean out the border sanctuaries of the NLF. South Vietnamese troops ranged deep into the country, however, committing terrible atrocities against Khmer civilians.[15] Sihanouk, in Peking, issued a call to all Kampucheans to join him and the previously tiny Khmer communist movement, once his enemies whom he had derisively labeled the Khmer Rouge (Red Khmer), in a massive, broad-based coalition to resist and overthrow the U.S.-backed Lon Nol government. Thus, the Lon Nol coup brought tensions in Kampuchea to the point of violence and plunged Kampuchea into civil war. The suffering of the people of Kampuchea has continued ceaselessly since March 1970.

So radical was the Khmer Rouge revolution that it has obscured the gravity of the situation in Kampuchea during the period of civil war between the U.S.-backed Lon Nol government and the Khmer Rouge from 1970 to 1975. This perspective is fostered by the political line of the present Heng Samrin government, which places primary responsibility for the destruction of Kampuchea on the "Chinese-backed Pol Pot clique." Countless foreign visitors and aid workers since 1979 have been shown ruined pagodas, vandalized machinery, and uncounted damaged shells of concrete buildings and have been informed that these were destroyed "by Pol Pot."[16] Many Kampucheans, including peasants, speak in terms of "before 75" and "after 75" as if the former time were the days of paradise. Given the destruction of Kampuchea during the "before 75" civil war period, this is eloquent testimony to the impact of the Pol Pot time on Khmer consciousness.

Statistics from the civil war period are numbing. In 1973, U.S. B-52s dropped 40,000 tons of bombs *per month* from early February until Congress forced a halt on August 15th. All pretense that the United States was bombing isolated communist enclaves was dropped; the area bombed cut a wide swath through some of the most productive rice land in Kampuchea.[17] B-52s bombed the rubber plantations in Kompong Cham, destroying plantations which in peacetime were an important source of foreign exchange. Draft animals by the hundreds were killed during the bombing; it is estimated that by the end of the war 75 percent of all domestic animals were killed.[18] In the ebb and flow of the fighting in rural areas, concrete buildings were used as cover for troops of both sides, resulting in war-related damage to hospitals, schools and pagodas. Up to one half of the hospitals in Kampuchea were destroyed in the civil war. Doctors in the embattled provincial capitals either fled to Phnom Penh or left the country altogether. Medical care in the countryside, never developed even in peacetime, broke down completely.

By 1974 little of Kampuchea's peacetime economy was functioning. The amount of rice land in the control of the Lon Nol government was approximately 500,000 hectares, compared to the normal figure of between 2 and 2.5 million.[19] While peasants were cultivating some rice land in zones controlled by the Khmer Rouge, most of Kampuchea's rice plain was a contested area where warfare prevented planting. Kampuchea, once a rice exporter, had to import 280,000 tons of rice from the United States under the Food for Peace program in 1974 alone.[20] To put this figure in perspective, this tonnage is equal to the total amount of World Food Program (WFP) food shipments during the entire post-1979 emergency and recovery period through the end of 1982.[21] Furthermore, the food shipped to Kampuchea was delivered to the Lon Nol government authorities *for sale* to the people in desperate need of food relief. Sales from food aid went either to the government to support its military campaign against the Khmer Rouge or into the pockets of corrupt officials. Out of $72.5 million spent on food aid for Kampuchea, only $1 million was used for free distribution to hungry people.[22]

In the last months of the Lon Nol government, the living conditions in government-controlled areas deteriorated completely. The population of Phnom Penh, a city of 600,000 in peacetime, swelled to about 2.5 million as refugees poured into the city to flee the bombing, the war, and the Khmer Rouge. With much of the rice-producing countryside in Khmer Rouge hands, this population was entirely dependent on outside food shipments for survival. Initially this food came up the Mekong River by barge, but when the Khmer Rouge cut the Mekong route in February 1975, the United States was forced to fly 40 DC-8s and C-130s into Phnom Penh daily to feed the population.[23] Rations plunged to 8.25 kilograms of rice per month, almost four kg below the amount the World Health Organization considers a minimum.[24] In some of Phnom Penh's refugee centers, the rate of severe malnutrition among children was 31 percent. The medical director of Catholic Relief Services declared in March 1975 that "hundreds" were dying of malnutrition every day. A doctor for World Vision stated, "This generation is going to be a lost generation of children. Malnutrition is going to affect their numbers and their mental capacities. So as well as knocking off a generation of young men, the war is knocking off a generation of children."[25]

THE KHMER ROUGE PERIOD

The exhausted Khmer Rouge peasant army marched into Phnom Penh on April 17, 1975, and its leadership assumed responsibility for the massive humanitarian problems facing the populace. The choice of this leadership, taken at least ten days before the fall of the capital and in keeping with its policies in previously liberated zones, was to evacuate the entire population of the city. The Khmer Rouge cadre and soldiers implementing the plan told the residents that it was only a temporary expedient to protect the population (and the revolution) from bombardments by the United States. In fact, the emptying of the city was permanent. With the exception of about 10,000 workers left behind to maintain production at some of Phnom Penh's factories, everyone was immediately sent out to the rural areas to work intensively on reviving Kampuchea's food production.[26]

The Khmer Rouge leadership, and those sympathetic to the revolution, have justified the evacuation on the following grounds:

a) security concerns dictated an immediate evacuation to prevent sabotage of the revolution by those loyal to the Lon Nol regime, or by outside intervention in some form by the United States;

b) the food and health situation in the capital was so precarious that people had to be evacuated at once to reach rural food supplies, avoid epidemics, and begin the urgent task of planting rice to ensure future self-sufficiency;

c) the Khmer Rouge had few cadre able to administer a city the size of Phnom Penh; purely as a matter of administrative efficiency and control it made more sense to break the population down into smaller units outside of the urban centers so alien to the cadre.[27]

These justifications, all of which contain elements of truth about the situation on April 17, 1975, cannot obscure how radical the Khmer Rouge decision to evacuate the urban areas actually was. The evacuation and the virtual elimination of urban life over the next three years and nine months of Khmer Rouge rule were unprecedented. These justifications avoid entirely the ideological underpinnings of the Pol Pot movement. This movement featured a utopian belief in the dignity and strength of the poorest peasants coupled with a fierce, paranoid, xenophobic Khmer nationalism which harked back to the glories of the great Angkor Wat civilization. The corrupt, foreign-dominated urban centers had to, and in fact did, cease to exist. The residents of these cities—Sino-Khmer merchants, ethnic Vietnamese, and educated Khmer infatuated with Western values and lifestyles—had to learn what it meant to be the purest, most honorable Kampuchean of them all—the poor peasant. This complete social revolution could only be accomplished by evacuating Phnom Penh and ending urban life.

The humanitarian justification ignores the fact that the Khmer Rouge actually turned down offers of aid from the French government and from UNICEF in the immediate aftermath of their victory.[28] With hindsight it is easy to see that the Khmer Rouge wanted no part of the normal Western aid system, with its requirements for imported technology, expatriate administrators, monitoring visits, etc. In this context, the immediate freezing of Kampuchean assets in the United States and the imposition of the trade embargo by the U.S. government had little significance. Indeed, to show their contempt for money and its uses the Khmer Rouge blew up the National Bank just after capturing Phnom Penh. Thus, U.S. aid policies in the aftermath of the Khmer Rouge victory had little relevance to the approach of the revolutionary leadership to solving the problems they inherited in 1975.

In March 1976, Radio Phnom Penh described Democratic Kampuchea, the official name of the country ruled by the Khmer Rouge, as "one huge worksite; wherever one may be, something is being built."[29] The workers on this huge worksite were divided into two broad categories: the "old" people, or those who had supported or been liberated by the Khmer Rouge well before the final victory; and the "new" people, or those liberated on April 17, 1975. The old people, among whom numbered many of the poorest peasants in Kampuchea, were the privileged group in Democratic Kampuchea. They lived in their own houses, received higher rations, and had more freedom to forage for food. They formed the village-level cadre and the peasant army of the Khmer Rouge. This power meant access to food which in some areas was systematically denied to the new people. The latter, including both urban dwellers and peasants who had fled to the cities, were given lower rations and were confined to the settlements. In some areas, particularly Battambang province in the northwest, they were worked very hard on starvation rations as a form of punishment for their former lives as city people.[30] Work was unrelenting for both groups throughout the Pol Pot period, but starvation, overwork and disease particularly decimated the former urban dwellers.

Since 1979 the urban dwellers able to speak English or French have poured out the stories of their suffering under the Khmer Rouge to international aid workers. They have emphasized the length and hardship of the physical labor they were forced to do and the pitiful food rations they received to sustain themselves. Many watched loved ones—parents, spouses, children—starve to death before their eyes as they stood by helplessly. One man, who has survived to work in the Hydrology Department of the Heng Samrin government, watched his father deteriorate for several months. When the father realized he was dying, he made a final request to his son—to bring him a bowl of rice. His son could not comply with this request.

It is a terrible irony in this context that the goal of the Khmer Rouge was to revive and expand rice production as rapidly as possible. This would be accomplished by abandoning unproductive land, expanding into previously uncultivated areas, raising yields through intensive use of natural fertilizers, and, most importantly, vastly expanding irrigation to match the glories of the Angkor era irrigation systems. Much of the heavy labor during the Pol Pot period involved shifting earth using only simple hand tools to construct new canals, dikes, and dams. Although few of these systems, many of which were built in isolated areas now abandoned, are still functioning today, the physical evidence of the number and size of these projects, and the mass mobilization needed to carry them out, is staggering.

One of the exaggerations about the Pol Pot period, fostered by the current government in Phnom Penh, is that all modern machinery was destroyed in an attempt to bring an absolute end to dependence on imported technology. As just noted above, the bulk of the heavy labor for the irrigation projects was done by hand; yet, tractors were used in Battambang, the traditional rice bowl province of Kampuchea whose vast, flat fields are one area in the country suited to mechanized agriculture.[31] As time went on, however, shortages of spare parts for Lon Nol era equipment probably became acute. In 1977, with the help of the Chinese, Democratic Kampuchea placed a $1 million order for Massey-Ferguson spare parts in Singapore.[32] The Chinese also supplied new tractors. Even with this assistance, yields in the floating rice areas of Phnom Srok district in Battambang were lower during the Khmer Rouge period due to the lack of tractors to turn over the soil to adequate depths.[33]

Despite the Khmer Rouge emphasis on self-reliance, the regime grew more not less dependent on China during the course of Pol Pot's rule. China provided not only tractors, but rice, cloth, drugs, and fuel.[34] Turnkey projects built with Chinese assistance included a textile factory in Kompong Cham and a phosphate factory in Kampot. The latter project included a small housing development for the Chinese advisors on construction and operation of the plant. Just before the Vietnamese invasion in December 1978, there were 20,000 Chinese advisors in the country.[35] This is probably three or four times the number of Vietnamese advisors in Kampuchea today. As the border skirmishes with Vietnam threatened to escalate into full-scale conflict, and as the harsh policies of Pol Pot and his followers began to provoke repeated internal rebellions, Democratic Kampuchea "could have been forced to rely so extensively on Chinese assistance that Kampuchea would have effectively lost any ability to act independently of China."[36]

Yet the Pol Pot regime did succeed in rapidly expanding rice production. Although no exact figures are available, information gathered from interviews with refugees in Thailand and conversations with people inside the country since 1979 suggests that in general rice production was quite good. During 1975 people had to cope not only with the legacy of the civil war, but with inefficiencies involved in organizing an entirely new structure for society. During this "apprenticeship," city people had to learn how to do manual labor, peasants how to work communally, and cadres how to organize thousands of people.[37] Yet the 1975-76 rice crop was excellent.[38] By 1977 and 1978, according to some estimates, the pre-war figure of 2.5 million hectares of rice land had again been obtained.[39]

The problem in Democratic Kampuchea was not food production but access to food. In some parts of Kampuchea, notably the East and Southwest zones of the Khmer Rouge administration, even new people had reasonable access to the rice they were producing.

In the Northwest, however, ironically the traditional rice bowl of Kampuchea, new people were provided with starvation rations. Many were forced to work in isolated forest areas where rice had never been cultivated and where virulent strains of malaria were rampant. Further, inexperienced and insecure cadre extracted too much food from their cooperatives or districts for shipment to Phnom Penh to create the illusion with the leadership that they were meeting quotas which had been set unreasonably high by the central authorities.

By 1977, the threat of hostilities with Vietnam led the center to force a greater and greater surplus from the countryside to stockpile in preparation for war.[40] Malnourished and starving workers harvested abundant rice crops in 1977-78 only to have the rice taken away by truck or by boat. This exacerbated the inequities in the distribution system based on class divisions from the pre-1975 period. Deaths from hunger or from illness made more acute by hunger increased markedly in all areas in the latter part of the Pol Pot period.[41] At the same time purge followed purge as factions within the Khmer Rouge leadership recognized the disastrous turn the revolution had taken and rebelled. Pol Pot emerged from these struggles as the dominant leader; he did nothing to change the course of the Khmer Rouge revolution. But these struggles further weakened the country at the very moment the xenophobic Pol Pot group was dreaming of full-scale war with Vietnam in late 1977 and 1978.

The human cost of the three-year and nine-month social experiment of the Khmer Rouge has been estimated at anywhere from 500 thousand to 3 million people. That recent careful estimates place total deaths towards the lower end of this scale (approximately 1 million) does nothing to vindicate or justify the policies of the Pol Pot government.[42] These policies had the effect of prolonging the humanitarian disaster of the 1970s for much of the population, despite impressive gains in total food production. The fanatic nationalism of Pol Pot and his followers led them to believe that this nation of 6 million people, crippled by war and starvation, could support a full-scale war against the battle-hardened Vietnamese. When Vietnam invaded Kampuchea in December, 1978 in support of a small group of Kampuchean dissidents and their followers, the collapse of the Pol Pot regime was sudden and complete.

THE 1979 FAMINE
AND THE INTERNATIONAL RESPONSE

The Vietnamese invasion liberated the Khmer people from the oppressive Pol Pot regime. The invasion ushered in a period of chaos, however, which resulted in the disruption of rice production and wide-scale famine, prolonging the nearly continuous food problems experienced by the majority of the rural population since the onset of the civil war in 1970.

It requires little imagination to picture the suffering of the Khmer people in early 1979. Exhausted and malnourished survivors, liberated with shocking suddenness by their traditional enemies, began to trek across Kampuchea to return to their homes and to search for missing relatives. As they walked, fighting between Pol Pot and Vietnamese forces swirled around them. Recapture by Khmer Rouge forces meant instant death for some or, at best, forced march west with the retreating troops. The Khmer Rouge pursued a scorched earth policy, destroying what they could not carry with them in their flight. The Vietnamese had attacked during the harvest, so in areas where the harvest could not be completed under Khmer Rouge control, the retreating troops mined the rice fields.[43] In areas controlled by Vietnamese and Kampuchean forces loyal to Heng Samrin, the leader whom the Vietnamese placed in power in Phnom Penh, hungry people looted stores and granaries, eating their fill of rice for the first time in months. No effective organization existed to protect or ration these stocks, which would soon be desperately needed.

As much as one half the population of Kampuchea migrated during 1979.[44] Some survivors, wanting no part of the Vietnamese or of yet another communist regime, fled immediately to Thailand. Many set off in search of relatives and friends. The trek of the survivors across the length and breadth of Kampuchea took many weeks. When people reached their villages they often found nothing left: no houses, the wat (pagoda) destroyed, not a trace of family and friends. Draft animals roamed the countryside, there for the taking, but there was little rice available for food, much less for seed for the main rainy season rice crop. Peasants devoted what little energy they had to scavenging for food and constructing shelter for their families. In the words of a village leader from Kbal Damrei in Pursat province, "In 1979, when we returned to our village, we had nothing, absolutely nothing."[45]

For the first six months of 1979, as the Heng Samrin government struggled to establish a working administration from the survivors of the holocaust, the only aid reaching Kampuchea was limited amounts of food from Vietnam and the Soviet Union. The magnitude of the crisis became clear to international aid agencies from accounts of the first refugees to reach Thailand, satellite photos which showed that virtually no planting had taken place for the 1979 rainy season rice crop, and a few independent eyewitness accounts from inside the country.

Not until July, however, were representatives from western aid organizations able to gain access to Kampuchea to assess the situation firsthand. Jim Howard, OXFAM's Technical Advisor with years of experience in disaster relief, flew into Kampuchea on August 26th with two doctors from the French Medical Aid Committee on the first relief flight of any significance to reach Phnom Penh. During a stay of nearly two weeks, he was able to travel outside Phnom Penh, observe conditions in the country, and talk to government officials and other survivors. During the first few days he travelled to the northern outskirts of the city where people were being held to prevent an uncontrolled return to Phnom Penh. Howard reported, "Visited small clinic at Kilometer 7, absolutely no drugs or medicines—serious cases of starvation—clearly just dying for lack of food. . . . The hundreds of children seen were all marasmic— much skin disease—baldness—discolored hair and great fear in the whole population." Beyond Kilometer 7 on Highway 5 "hundreds of people —were on the move pushing or carrying their few belongings towards Phnom Penh."[46]

A few days later Howard visited Kompong Speu province, about 40 km south of Phnom Penh. The town had been a Pol Pot stronghold and in February retreating Khmer Rouge troops had ignited an ammunition dump on the site of the former central market. "The town is completely destroyed and there are hundreds of tons of live grenades—shell cases—and heavy calibre amunition laying around the site." Outside the town itself, Howard reported, "the villages visited all contained starving people and clearly many of the people I saw couldn't possibly survive several more months on what they had available. Most had a tiny rice ration of 3 kg per month—and they were eating wild tree pods and cooking banana stems. This was starvation at the worst Biafra level—much due to able bodied men being absent or dead—and, therefore, the villages had no work capacity."[47]

The gravity of the emergency facing the people of Kampuchea overwhelmed Howard and other early visitors to Phnom Penh. For not only were food and medicine in short supply, but very little planting had been done for the 1979 monsoon crop, which normally supplies 90 percent of Kampuchea's rice needs. Thus, internal sources of food in significant quantities were not going to be available through 1979 and into 1980. Howard predicted that two million Khmer, then believed to be half the surviving population, would die in three months without an immediate and massive humanitarian response.[48] While in hindsight it is possible Howard unintentionally exaggerated the crisis (and underestimated the value of foods a starving person could scavenge in the countryside), his emotional testimony galvanized the international community and created the urgency needed to mount the required response to the disaster.

Given that President Carter had labelled the Pol Pot government "the worst violator of human rights in the world" in 1978, the United States should have been prepared for the disaster when that government was overthrown. The fact that it was the Vietnamese who precipitated the collapse, however, complicated the U.S. response to the crisis. Despite dire reports on the internal situation coming from the U.S. Embassy in Bangkok, State Department officials in Washington considered the so-called famine to be a propoganda ploy by the Vietnamese to get aid and *de facto* recognition for the Heng Samrin government, which they had placed in power. Even a personal visit to Washington by the U.S. Ambassador to Thailand, Morton Abramowitz, failed to convince State Department officials that the famine was real. Only when satellite photographs showed that only 10 percent of Kampuchea's rice land was under cultivation did Washington-based officials begin to take the emergency seriously.[49] Administrative obstacles related to negotiating an international aid presence in Phnom Penh kept U.S. aid at token levels to the Thai border refugee encampments through September.

Ironically, what tipped the balance in favor of mounting perhaps the largest international relief effort ever was the spectacle of battered Khmer Rouge cadre and their families who entered Thailand *en masse* under heavy Vietnamese military pressure in September and October. Many of these people had been on the run for weeks, wandering in forest areas where food supplies were unavailable. When these obviously suffering people finally crossed the border and collapsed into Thailand, their appearance shocked the world and forced governments to take action to "save the refugees." That these refugees just one year before had been the village-level cadre for the worst human rights violator in the world got lost amid painful images of starving women and children.

Yet as pressure from individuals and governments mounted, the aid organizations still had to tiptoe their way through a political minefield to bring aid to Khmer people in need. The Heng Samrin government, stung by the lack of recognition accorded it and still engaged in subduing Pol Pot resistance, insisted that all aid be channeled through Phnom Penh. The authorities would tolerate nothing that would violate the sovereignty and national interest of the People's Republic of Kampuchea (PRK). Relief aid at the border to the Pol Pot remnants and other refugees challenged the legitimacy of the government and set up a tiny "alternative" Kampuchea just inside Thailand. Thus, agencies which wished to work inside Kampuchea had to renounce aiding Kampucheans who were outside the control of the Heng Samrin authorities. Western governments and media bitterly criticized the Heng Samrin government for what

Debris litters a main street in Phnom
Penh, September 1979. In the immediate aftermath of both the Khmer
Rouge triumph in 1975 and the Vietnamese invasion in 1979, shops and
homes were looted in search of items
of value; garbage and smashed house-
hold goods were left in piles in the
streets, by Marcus Thompson
OXFAM.

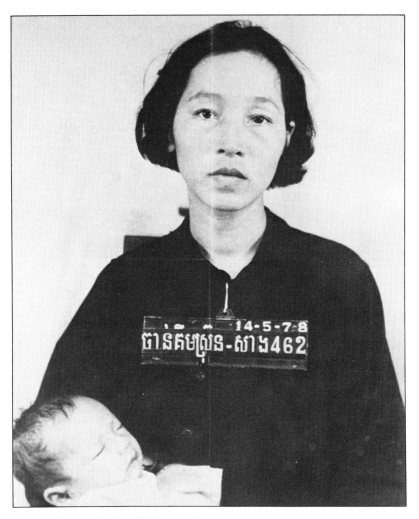

Two victims of Pol Pot killed in Tuol
Sleng prison in Phnom Penh in 1978.
Their photograph, taken by their
killers, adorns the walls of the prison
which is now a museum and memorial
devoted to the victims of Khmer Rouge
rule, by Laurence Simon
Oxfam America.

One of the many displaced persons who crossed Kampuchea to return to their native villages in 1979, by Don Luce.

Orphans weave mats at Orphanage No. 1 in Phnom Penh, by Laurence Simon Oxfam America.

First-year class at Chaktamouk
School studies writing, by John
Spragens, Jr.

Worker weighs bags of phosphate
at the Battambang Phosphate Factory
supported with assistance by Oxfam
America for spare parts, plastic sacks,
and transport equipment, by Joel
Charny/Oxfam America.

Peasants in Prey Veng
gather rice plants
for transplanting
in nearby paddies,
by Laurence Simon
Oxfam America.

Production solidarity group
threshes rice in
Kompong Chhnang,
by John Spragens, Jr.

Peasant woman in Prey Veng irrigates her fields using the *rohat*, a wooden, pedal-powered traditional pump used extensively in eastern Kampuchea, by Laurence Simon Oxfam America.

Harvesting rice in
Muong village,
Kompong Cham province,
by John Spragens, Jr.

was perceived as an inhumane, purely political stand which jeopardized the lives of thousands of Khmer people.[50]

OXFAM was the only international organization which accepted these conditions in an agreement worked out with Foreign Minister Hun Sen in early October, 1979. Eventually many agencies, including UNICEF, International Committee of Red Cross (ICRC), World Vision, AFSC and Church World Service, were able to mount significant programs on both sides of the political conflict. OXFAM, however, by taking the bold step of agreeing to the PRK conditions, broke the political log jam and opened up the aid channel to Phnom Penh.

Upon reaching the agreement with Hun Sen, OXFAM dispatched Guy Stringer, its Deputy Director General, to Bangkok and Singapore to put together a barge of relief supplies to Kompong Som.[51] The obstacles to pioneering the route which eventually became one of Kampuchea's lifelines were formidable. The Thai government, for example, was blocking all shipments of relief goods purchased in Thailand for Kampuchea. In Singapore, the government was only slightly more open to Stringer's mission; officials were extremely wary of press coverage of a potential breach in ASEAN unity against Vietnamese-occupied Kampuchea. Singapore dock workers refused to load cargo for Kampuchea until they received personal assurances from OXFAM staff that the relief goods would not be diverted to Vietnamese soldiers. Finally, there was no certainty regarding security in Kompong Som port, or the capacity of the Khmer authorities to organize the unloading of the food and medicine which OXFAM was sending.

The barge, pulled by a tug and loaded with 1,500 tons of relief supplies, arrived at Kompong Som on October 14th after a five-day journey from Singapore. The barge was met by the Minister of Economy and Reconstruction, the Minister of Health, the Minister of Agriculture, and the provincial Governor. The Minister of Economy's direct management was required to organize the dock workers to unload the cargo and trucks to transport the goods to Phnom Penh. Virtually every working truck in Kampuchea at that time, little more than 100, greeted the OXFAM barge. It took nearly five days for the team of inexperienced dock workers to unload all the cargo.

Stringer spent the rest of his two-week visit organizing, along with other OXFAM staff, the next barges to Kampuchea. Already with the third barge, OXFAM was beginning to direct its program towards the reconstruction needs of the country. Barge number 3 carried short season rice seed, maize seed, hand tools, cotton yarn for a textile mill, fishing nets, nylon twine for the fishnet factory, and diesel pumps for farm-level irrigation—all this in addition to the desperately-needed rice, sugar, and edible oil for supplementary feeding programs. OXFAM also hit upon the idea of bringing barges up the Mekong to Phnom Penh, transporting aid closer to the central warehouses and distribution network of the Heng Samrin government. In November 1979, OXFAM brought the first Western barge up the Mekong to Phnom Penh since February 1975.

Political Obstacles to the Aid Effort

Even with this breakthrough, however, UNICEF and ICRC still had problems gaining acceptance by PRK officials for an international presence which would allow them to implement the massive aid program required by conditions inside the country. The government's deep suspicion of Western agencies was justified under the circumstances. Indeed, the very countries offering aid out of one hand had just finished raising the other in support of continued seating of Pol Pot as the legitimate representative of the people of Kampuchea to the United Nations. This weak and insecure government, composed of exiles to Vietnam, rebels against the Khmer Rouge, and hundreds of survivors of the Pol Pot holocaust, could not welcome representatives of the hostile outside world with open arms. The intransigence of the PRK, however, alienated the staff of Western aid organizations and reduced the amount of aid reaching Phnom Penh in the initial stages of the emergency. These political difficulties, coupled with severe logistical problems inside the country, meant that massive amounts of aid never did reach the Khmer people before the end of the year.

The political obstacles to aiding all Kampucheans in need plagued the relief effort. There were at least three distinct components of the overall international aid program: one concentrated on or near the border between Kampuchea and Thailand, where the food and medical requirements of refugees had to be met with no prospect for long-term self-sufficiency; one through the Heng Samrin government which controlled all of Kampuchea's rice land and which offered possibilities for long-term self-reliance for the 6 million survivors of the Pol Pot period; and finally, the effort based at the border to contribute to the recovery in the interior by distributing supplies to people who had come to the border camps with the intention of returning to their villages. The "landbridge," as this latter component was called, was initiated at a later stage in the emergency period in response to perceived inadequacies of the channel through Phnom Penh.

Each of these components had a political dimension ripe for exploitation by governments and political factions. The border encampments in Thailand became havens for retreating Khmer

Rouge soldiers and their families, in addition to elements of the fledgling, poorly organized non-communist resistance to the Vietnamese occupation. The extent to which Khmer refugees were immediately fed and cared for by international agencies working at the border would determine the long-term viability of the resistance. From the beginning it was impossible to keep food and medical aid at the border from going to the soldiers of the various factions. Indeed, at various stages in the chaotic late 1979-early 1980 period, the refugee camps at the border were administered directly and openly by military men who controlled the distribution systems for humanitarian aid.[52] Thus, a key aspect of the border program was building up the resistance, which served the strategic interests of Thailand, China and the United States. And since the strongest element by far in the resistance was (and is to this day) Pol Pot's Khmer Rouge, the border relief program contributed to saving and strengthening the movement responsible for the deaths of more than one million Kampucheans.[53]

Aid to the interior had an equally significant political impact. The Heng Samrin government insisted from the beginning that all aid be channelled through its institutions. Unlike the border, there could be no direct distribution, and the large foreign presence needed to implement such schemes could not be tolerated. By insisting on control of the humanitarian aid flowing into the country, the Heng Samrin government was able to use this aid to strengthen its terribly weak administration and gain credibility with the rather skeptical populace. There is nothing unusual about this; governments routinely use humanitarian aid in support of quite narrow political objectives, as in the cases of the famine in Biafra and that of Somoza-ruled Nicaragua after the 1972 earthquake. What differentiates Kampuchea in 1979 is that the vast majority of the world's nations, including the major government donors to the United Nations relief program, did not recognize the legitimacy of the government which controlled the territory and population of the country. To aid the six million survivors inside Kampuchea, donor countries essentially had to agree that this task was noble enough to overcome their distaste for aiding a Vietnamese-installed government. That they chose to do so allowed the Heng Samrin government to consolidate its control over the territory of Kampuchea.

The cross-border program grew directly out of these contradictions. The leading donor countries, particularly the United States, were actively supporting military resistance to the Vietnamese occupation forces and political destabilization of the Heng Samrin government, even as they sent emergency aid to this very government. In this context, the landbridge program served a dual political purpose: it kept humanitarian aid out of the distribution networks of the Heng Samrin authorities and it served to destabilize the interior by bringing people to the border who would otherwise remain in their villages in Kampuchea. The more people going back and forth to the border, the more resistance forces could slip back into the interior along with peasants with rice seed, smugglers, etc. That the Vietnamese considered this a political/strategic program rather than a humanitarian one was shown by their military attack on the major cross-border distribution point at Nong Chan on June 23, 1980. By this time, of course, thousands of peasants had already traveled back and forth to the border in search of rice seed, hand tools, food, and consumer goods for their families.[54]

Essentially there was a polarity to the relief effort: an interior program administered by Heng Samrin government officials, with monitoring by the personnel of the international agencies based in Phnom Penh; and the border effort with much greater administrative involvement by the international agencies and Western donor governments. While the United States gave generously to both programs, its political interests in undermining the Heng Samrin government conflicted with its interest in seeing humanitarian aid reach the majority of Kampuchea's population. Thus, even as the United States contributed to the relief effort on both sides of the conflict, its Embassy in Bangkok did everything in its power to discredit the People's Republic of Kampuchea and its administration of aid to the interior. According to the U.S. interpretation, the stockpiling of food in Phnom Penh's warehouses in late 1979 represented willful denial of humanitarian aid, not the fact that there were less than 200 trucks in the entire country.[55] Inefficiencies due to the inexperience of PRK cadre were interpreted in a similar fashion. OXFAM was severely criticized for being a dupe of the Vietnamese, and the U.S. Embassy in Bangkok later spread a false rumor that rice seed sent into Kampuchea by the OXFAM/Non-Governmental Organization (NGO) Consortium had been treated with a deadly mercury-based fungicide which would poison Khmer peasants handling or eating the seed.[56] In a separate report the CIA blamed the Heng Samrin regime for 600,000 deaths from famine in 1979 and warned that even more deaths from starvation were in the offing unless 450,000 tons of food could be distributed to the Khmer people before July 1, 1980.[57] This report, the stream of negative information — really deliberate disinformation in the light of events — from U.S. Embassy staff, and judgments of other strongly anti-Vietnamese sources at the border led even normally acute observers like William Shawcross, author of *Sideshow*, to speculate that the Khmer people were on the verge of extinction. The Vietnamese, therefore, not Pol Pot, would be ultimately responsible for "genocide" in Kampuchea.[58]

The Relief Effort

Taken out of this intense, polarized political context, a logical scenario for the recovery of the Khmer people in 1979-80 can now be reconstructed. The key event of 1979, which allowed the massive international relief effort to begin, was the establishment of security for the people of Kampuchea by the Vietnamese army in late 1979. Until then, as long as Pol Pot remnants could execute anyone under Heng Samrin control as a traitor to the Khmer people, peasants lacked the security necessary to begin reviving the food production system. Once it was established that a foraging trip for fruit or wild food sources or fish would not risk one's life, the survivors could begin to rely on the natural reserves of the Kampuchean food system. The ability to forage, when coupled with the relief effort which finally got underway on a large scale in early 1980, meant that the food needs of the bulk of the population could be met, albeit at a subsistence level resulting in a great deal of malnutrition and risk to vulnerable groups, particularly children. None of this would have been possible, however, had the resistance of Pol Pot continued at significant levels throughout the countryside into 1980.

By early 1980, therefore, enough stability prevailed inside the country to allow the people of Kampuchea assisted by governments and the international organizations to turn their attention to the revival of food production in Kampuchea; this was the crucial component of self-reliance for people who had been unable to control productive resources for almost a decade. The consequences of the human disaster heightened traditional agricultural problems which had plagued Khmer peasants for centuries: lack of irrigation facilities and low yields as a result of poor soils and of the use of traditional rice varieties without chemical or even natural fertilizers. In addition, the human and draft animal populations, formerly adequate to work enough land for rice production, had dropped drastically in the past decade. Kampuchea, never overpopulated, now lacked the human resources to feed itself. Even hand tools and wood plows were scarce. Finally, the overall food shortage meant that seed stocks, normally maintained from harvest to harvest, were either seriously low or depleted altogether.

In response to these problems, the international aid agencies and bilateral donors imported over 400,000 hoe heads, 5,000 small diesel irrigation pumps, 1,300 power tillers, 55,000 metric tons (mt) of rice seeds (most them international high-yielding varieties) and many thousands of tons of the chemical fertilizers essential to their proper growth and production.[59] This aid represented a massive influx of technology used only rarely by Khmer peasants before the disaster. While tractors had been used extensively in Battambang province before 1975, diesel irrigation pumps were scarce and employed primarily by wealthy farmers.[60] In 1967 imported international rice varieties constituted at most three percent of the total rice area of 2.51 million ha.[61] In contrast, the 42,000 tons of hybrid rice seed supplied for the 1980-81 planting season, if utilized properly, would have planted about 525,000 ha, or nearly 40 percent of the area cultivated. And the figure of 42,000 tons includes only seed sent through Phnom Penh, not the many additional tons of seed sent across the border as part of the landbridge program.

Table 1 summarizes the comparison between 1968/69 and 1980/81:

Table 1

	1968/69	1980/81
Rice cultivation 000 ha	2,427	1,320
Paddy production 000 mt	2,503	1,580
Diesel irrigation pumps	1,300	5,000
Tractors	1,500	200
Chemical fertilizer imports mt	5,000	29,800
Hybrid rice seed imports mt	not available	42,000

Sources: World Bank for 1968/69 data; FAO for 1980/81 data.

In effect, therefore, the relief effort supplied technology that Kampuchean rice farmers had never used before, even in times of stability, when they had barely emerged from a decade-long period of famine and destruction. They had to use these imported technologies for the rice crop upon which their survival depended. Yet, in a disaster situation peasants are most likely to rely on the customs and practices which have ensured survival of the people and their culture. This goes a long way towards explaining why despite the tons of imported rice seed which were inundating Kampuchea, it was practically impossible to find these varieties in farmers' fields in 1980. The risk was too great that something would go wrong: flooding of the short-stemmed hybrids in low-lying areas, lack of fuel for the diesel irrigation pumps essential to ensure adequate water supply to the varieties in rain-fed areas, shortages of fertilizers when application was required to ensure promised yields. High-yielding varieties function in a clockwork fashion and the absence of an input at each phase of the rice cycle seriously jeopardizes the yields at harvest time. Only when no other seed was available did it make sense for a Kampuchean peasant to increase the risk of losing his entire crop. If circumstances permitted, it was much better to eat the international seeds in anticipation of customary yields with traditional Khmer varieties.

That many Khmer peasants chose to eat rather than plant international seed varieties would help explain why, despite the dire predictions of many observers, the people of Kampuchea survived 1980 until the harvest without a recurrence of the famine conditions of 1979. In the first eight or nine months of 1980, before the World Food Program had its food purchasing and shipping logistics in order for the Phnom Penh operation, most of the rice reaching the countryside was in the form of seed not food. Eating this seed (seed rice is simply unhusked food rice) probably provided a crucial source of food. Equally important were the food resources of the Kampuchean countryside, resources which did not, in fact, depend on outside aid to be exploited. Fruit trees, vegetable gardens, and vast fishing reserves, from rice paddies and ponds to rivers and the Tonle Sap Lake provided essential supplements to the Khmer diet. During the 1980 rainy season it was virtually impossible to travel outside of Phnom Penh without seeing countless children fishing with poles from small dikes beside the rice paddies. Their families had given them the task of gathering needed protein for consumption. Manioc and corn, disdained by Khmer peasants as animal food in normal times, substituted for rice as staples. Again, the crucial factor was that villagers were once again free, as they had not been during the Pol Pot time, to exploit these valuable food sources. Villagers in Takeo province southeast of Phnom Penh told the OXFAM doctor in 1981 that the freedom to travel beyond their home villages to look for supplementary foods differentiated life under the current government from that under Pol Pot.[62]

The 1980 monsoon season rice planting yielded 1.46 million tons of paddy (unhusked rice) on 1.23 million ha, an average yield of 1.2 mt/ha.[63] This crop more than doubled the production for 1979 and represented a great achievement for the government and people of Kampuchea. While falling well short of creating a surplus, Kampuchea's rice farmers reached a subsistence level which they have been able to maintain in subsequent harvests, despite severe weather problems in 1981 and 1983.

Crops neither dependent on outside inputs nor on extensive

labor revived remarkably during the emergency period. By November, 1981 vegetable production had reached 1968/69 levels. Corn production peaked in 1980, but dropped off in 1981 due to flooding and the low prices paid to corn producers which led them to substitute more lucrative cash crops such as tobacco. In 1980 manioc production was more than ten times that of the 1960s, but as the food situation stabilized production declined to levels still significantly higher than normal.[64] Fish production still lagged at less than 50 percent of pre-war levels, but commercial fishing on the Tonle Sap lake is both labor intensive and dependent on resources not readily available locally (outboard motors, large fishing nets, and boats).[65]

The relief effort in the agricultural sector, and the revival in peasant production, could not have taken place without millions of dollars in assistance to the country's shattered infrastructure. The OXFAM/NGO Consortium and UNICEF, in addition to bilateral donors, supplied hundreds of trucks, river ferries, and barges which constituted the basis of Kampuchea's internal transportation and distribution network. The OXFAM Consortium led the international effort to repair Kampuchea's industries, concentrating on consumer-oriented factories in Phnom Penh, particularly textiles. The Consortium was also instrumental in supporting the efforts of Khmer technicians to assure adequate power and clean water to the workers and residents of the capital. In the countryside, rural clinics and provincial hospitals were staffed by medical teams primarily from the socialist countries, using medicines and equipment donated by ICRC and numerous private agencies of which the World Council of Churches was perhaps the largest donor. While most of the 200,000 children orphaned during the Pol Pot period lived with their relatives or even strangers who generously adopted them early in 1979, hundreds of others received assistance through orphanages established in towns and smaller district centers. International organizations, led by UNICEF, ICRC, and World Vision, provided beds, school supplies, medicines, and other items to these children's centers. This aid, in total, provided a basis for societal recovery of which agricultural production was perhaps the most important aspect.

Given the multitude of organizations working in Kampuchea, some of which were also mounting programs at the border, only a rough order of magnitude can be established for the total dollar value of the relief effort during the emergency period from August 1979 to December 1980. The joint U.N.-ICRC program coordinated by UNICEF and ICRC spent more than $250 million inside Kampuchea out of a total program of $489 million, which includes aid to border camps, refugee camps inside Thailand, and so-called "affected

Thais" in villages along the border.[66] The OXFAM/NGO Consortium mounted a $40 million program with contributions from both European and American agencies. Other major consortia included Acton for Relief and Rehabiliation in Kampuchea (ARRK), led by Church World Service, which spent more than $5 million and CIDSE (International Cooperation for Socio-economic Development), a group of European Catholic agencies which spent between $10-15 million. These programs, added to those of individual agencies such as AFSC, World Vision, and the World Council of Churches, yield an estimate of $70-80 million for total private voluntary aid during the emergency period.

To the Joint Mission program of $489 million the United States contributed about $137 million, more than a quarter of the total.[67] The Joint Mission report of its activities during the emergency period does not break down donations by program (interior vs. border), but by agency. The bulk of U.S. contributions ($71.6 million) went to the World Food Program in the form of surplus rice or cash to purchase stocks on the world market. The United States donated $31.4 million to the United Nations High Commission for Refugees for food and health supplies for Khmer refugees in Thailand. Even during the emergency period the United States gave only $5 million to FAO, indicating the unwillingness of the United States to contribute even to the short-term recovery of Kampuchean agriculture. Some U.S. funds supported the agricultural rehabilitation programs of American private humanitarian agencies, particularly the purchase of rice seed. Thus, quite early on the U.S. government showed itself unwilling to fund the supply of materials which would have a measurable impact on the long-term recovery and development of the people of Kampuchea. This bias in the allocation of U.S. funds presaged the tightening of the trade embargo during the post-famine period since 1981.

By early 1981, the threat of famine on a massive scale had ended in Kampuchea. The euphoria prevailing in Phnom Penh on January 7, 1981, the second anniversary of the liberation of Kampuchea from the Khmer Rouge, reflected the joy and wonder that the Khmer people felt at realizing they had survived yet another trial by fire. The city, which only 15 months before had been described by OXFAM Director General Brian Walker as a "silent city almost devoid of life,"[68] now sparkled with people celebrating not only the holiday but their own survival. The joy they experienced on that day has since given way to doubt and anxiety as the intractability of the long-term problems which they face has become clearer. The survivors had much to celebrate in January, 1981. How much they owed their endurance to the international relief effort and how much to their own skills, strength and resilience cannot possibly be determined.

AFTER THE FAMINE:
OBSTACLES TO RECOVERY

The revival of the food production systems of Kampuchea is the key to the reconstruction and future development prospects of this battered country. The tragic irony of a once food-exporting country experiencing wide-scale famine suggested this focus immediately during the emergency period in 1980. Rice is not only the staple of the Khmer diet, but as a commodity was also the country's major foreign exchange earner in the pre-war, independence period. Rice garnered as much as 60 percent of export revenues in bumper crop years.[69] In the long term, the revival of Kampuchea's food export capacity, assuming the existence of formerly reliable markets for its products, will provide the impetus for growth and development of the economy.

It was in this context, of a stablized but still deficit food situation, with long-term needs unmet, that the Department of State issued its February 2, 1981 memorandum on Kampuchean assistance analyzed in detail in Chapter I, above. The policy statement specifically rules out the use of U.S. funds for rehabilitation or development projects, citing agricultural aid "which would lead to surpluses" as an example.[70] As Kampuchea had reached a point where such aid was crucial to ensure further progress, the U.S. prohibition on development assistance has had the effect of nipping the country's recovery in the bud and prolonging the food deficit situation. A period of quiet but constant crisis has continued to the present.

Because the Kampuchean representation to the United Nations remained in the hands of the Khmer Rouge, U.N. agencies working in Kampuchea did so under a special mandate renewed annually by the U.N. General Assembly. This mandate was but one aspect of a political document whose primary purpose was to condemn the Vietnamese invasion and call for a withdrawal of all foreign troops from Kampuchea. The Kampuchea Resolution has received overwhelming support from member nations since 1979, with the United States, China, and the countries of the Association of Southeast Asian Nations (ASEAN), including Thailand, Singapore, Philippines, Malaysia, and Indonesia, leading the lobbying efforts on its behalf. The clause regarding U.N. aid has called for "humanitarian assistance" for Kampuchean people in need. This has been interpreted from the beginning of the relief effort as expressly prohibiting

U.N. agencies from providing long-term development assistance to the Heng Samrin government. Thus, the U.N. agency responsible for development assistance, the United Nations Development Program (UNDP), has been unable to work in Kampuchea at all since 1979. The U.N. Food and Agriculture Organization (FAO) has mounted programs which had to meet the criteria for emergency assistance as defined in the U.S. policy statement cited above. These restrictions forced FAO to limit itself to providing basic inputs almost exclusively: fertilizer, pesticides, rice seed, hoe heads, manual and motorized sprayers, small rice mills, animal vaccines, tractor spare parts, and hand tools.

While the quantities are impressive (53,720 mt of rice seed and 85,633 mt of fertilizer alone imported from 1979 through 1982), most of the aid provided was for one-time use and had only short-term impact.[71] Further, to get permission to provide even these basic supplies judged essential by FAO specialists, FAO had to file reports which exaggerated the negative aspects of the Kampuchea situation to justify continued "emergency" assistance. This damaged the credibility of FAO's Kampuchea program, even when their reporting proved accurate. Donor support for the special FAO program completely dried up in early 1983. In July 1983, in a country desperately needing long-term agricultural assistance, FAO put its vehicles and office equipment in Phnom Penh up for auction. The special program, as limited as it was, is without funding although FAO sent another food assessment mission to the country in late January 1984.

Even in the absence of the statutory limitations under which FAO was operating, the private voluntary agencies tended to follow an identical formula for increasing Kampuchean rice production: international high-yielding rice seed varieties plus imported chemical fertilizers equals greater food production. Apart from the high cost of purchasing and shipping these inputs, a liability in normal times but less so during the Kampuchea emergency, this formula was unsatisfactory for two reasons: the unsuitability of the international varieties to Kampuchean conditions, noted above, and the dependence on outside aid fostered by this approach. In return for the massive investment in agricultural inputs during the emergency, the rice farmers of Kampuchea gained almost nothing which would have a measurable impact on their lives over the next 10 or 20 years.

A strategy for making a contribution to food security in Kampuchea over the long term would concentrate on the following areas: improving labor productivity and efficiency, improving the quality of rice seed stocks, increasing the availability of local fertilizers, expanding irrigation, and creating local training institutions.

The extent to which these tasks are undertaken by the PRK authorities with outside assistance from international organizations and bilateral donors will determine when Kampuchea will obtain food self-reliance and security.

Labor Power

The huge increase in land cultivated with rice from 1979 to 1980 seemed to promise rapid expansion of hectarage in subsequent plantings. In fact, monsoon season rice cultivation has been more or less frozen at 1.1 - 1.3 million ha through 1983, with the actual size of the crop being determined by climatic factors. In 1981, the Mekong basin flooded to the highest level in one hundred years of records, while simultaneously many of Kampuchea's central rice-growing provinces experienced severe drought, causing losses of about 200,000 ha. Total paddy production dropped over 400,000 mt from 1980. While a reliable monsoon in 1982 meant lower losses, the hectarage of the 1982 planting again failed to surpass 1.3 million hectares against a goal of 1.5 million.[72] Drought in the eastern provinces and unprecedented flooding in Battambang province in the west resulted in signficant losses in 1983, showing once again how vulnerable Khmer peasants are to weather-related problems. Losses due to flooding in Battambang alone were more than 150,000 ha. Total hectarage again failed to surpass 1.5 million.[73]

This immediate stagnation of rice production results primarily from a severe labor shortage in the countryside. Population data for post-1970 Kampuchea are highly speculative, but most current estimates place the populaton at close to 7 million.[74] But as a legacy of the 1970s, the composition of the population is abnormally imbalanced: as little as 25 percent of the adult population is male, while 30 percent is under 15 years of age.[75] Further the draft animal population, decimated by war and disease, is only 1.24 million, or the level of the mid-1950s.[76] Thus, available labor in the countryside is about equal to the amount available to Khmer peasants in the 1950s, when they had to support a population of only 4.7 million people on about 1.1 million ha of rice land.[77] Given the shortage of labor to feed a population of nearly 7 million, Kampuchea, perhaps for the first time in its history, seems to have a problem of over-population, particularly with the explosion in births since 1979. Until these children grow up, *if* they survive to adolescence and adulthood, the peasants of Kampuchea will have a difficult time expanding production to keep up with the growth in population. The labor shortage in the countryside is further exacerbated by insecurity, which forces many young men to join the fledgling Heng

Samrin army rather than remain on the farm. This labor shortage results in a chronic production crisis which is a direct consequence of the disasters of the 1970s.

To protect the surviving draft animal population, FAO, with the assistance of several voluntary agencies, aided the Kampuchean Veterinary Department in conducting a mass vaccination program in 1981/82. The maintenance of a regular vaccination program over the next several years is essential to ensure that the draft animal population continues to grow. Breeding programs are beyond the capacity of the Veterinary Department and the Ministry of Agriculture at present, although there was consideration of starting a pilot ranch in late 1980 with breeding stock from India or elsewhere in Asia. Khmer peasants themselves do not breed draft animals.[78]

Whether as a result of government inefficiency in distribution and training or simple peasant resistance, peasants did not use the power tillers and iron buffalo imported as part of the relief effort. Occasionally peasants would hook up power tillers to a cart for local transport. Tractors, which were used extensively in western Kampuchea in the late 60s, would seem to have great potential in the land-rich provinces of Battambang and Siem Reap. The vastness of the western rice plains, and the relatively large size of the plots, cry out for mechanized exploitation as a means of expanding production. In Battambang there is even an excellent repair and maintenance facility which has managed to keep about 100 tractors operating in the province. The workers at the facility rehabilitated about 40 Massey Ferguson tractors, using primarily parts cannabalized from the many tractors destroyed in the civil war or by retreating Pol Pot troops. They showed such pride in their work that they even forged the distinctive "MF 135" plate to put on the front of the restored tractors. This worthy project, however, is one of several Oxfam America regretfully decided not to request a license to fund due to the unlikelihood of receiving such a license from the State Department. The former Director of this facility, a Berkeley-trained economist, estimated in early 1982 that an additional 100 tractors would be required to bring floating rice hectarage up to pre-war levels.[79]

The two major constraints on the full use of even the tiny tractor fleet in Kampuchea have been shortages of fuel and lack of trained maintenance personnel outside of Battambang province. Without adequate fuel supplies, which only bilateral donors such as the Soviet Union can assure at this point, and priority allocation of limited fuel resources to the agricultural sector, even a small-scale mechanization approach is doomed to failure. The importance of trained personnel is underscored by the 1982 experience with mechanization, when only Battambang province attained as much as 80 percent of the planned figure for mechanized land preparation; the country-wide figure was a dismal 34 percent.[80] Battambang's relative success can be attributed to much more extensive pre-war experience with mechanization and the existence of the well-organized repair facility just noted above.

Rice Seed

Therefore, use of improved seed stocks to increase yields on currently available land is extremely important. Khmer peasants routinely select the best seeds at harvest time, grouping them by variety for the next planting season. Choosing the varieties best adapted to the local soil and water conditions is the key task, but the taste of the rice is also an important consideration. A peasant may plant as many as five or six rice varieties on different small plots, thereby significantly lessening the risks of a total crop failure.[81]

The Kampuchean Ministry of Agriculture has selected IR 36, a variety developed by the International Rice Research Institute (IRRI) in the Philippines, as the international variety to promote. In actual farm conditions IR 36 produces about 3 mt/ha as opposed to only slightly more than 1 mt/ha for Kampuchean traditional varieties. But without fertilizers or water control, IR 36 yields no more than these traditional varieties and Khmer peasants much prefer the taste of the latter. Therefore, the pace of extension and acceptance of IR 36 has been slower than the production needs of Kampuchea might otherwise dictate. During the 1982 monsoon season, 66,000 ha of IR 36 were planted.[82] Even assuming excessive seeding at the rate of 100 kg per ha, the 6600 mt of IR 36 seed employed represents but a fraction of the total seed imported during the emergency period. Having dug themselves out from the avalanche of international seed, the Ministry of Agriculture and the peasants of Kampuchea are proceeding gradually with the expansion of high-yielding rice production.

The multiplication of traditional varieties is another top priority at the national level. Many pure traditional varieties were lost during the Pol Pot time due to extensive forced internal migrations of the peasantry. For example, in floating rice areas of Kompong Chhnang peasants complained that yields were only a fraction of those in the 1960s when they had the two rice varieties best suited to local conditions: *Kanlong phnom* ("jump over the hill") and *pok*. That rice seed varieties, one of *the* most crucial components of peasant agriculture, could actually be lost underscores the depth of the disruption of Khmer rural life under Pol Pot.

In late 1980, someone at the Ministry of Agriculture, probably Minister Kong Som Ol, remembered that IRRI had sent missions to Kampuchea before 1975 to collect seed varieties for its Germplasm

Bank, where over 60,000 of the world's rice varieties are stored in small quantities. IRRI has over 800 Khmer rice varieties in its storage facility; 140 of these have been returned to Kampuchea in the past two years under the auspices of Oxfam America. The Department of Agronomy's Research Division is presently multiplying these varieties at agricultural stations in Kompong Speu and Battambang. The Ministry's goal is to be able to produce these traditional varieties in sufficient quantities to distribute to peasants and keep on hand at the national level for future indigenous seed development programs.

Proper seed storage facilities are a necessity if the long-term goals of this and other seed storage programs are to be realized. While actual breeding of locally-adapted hybrids is a long-term prospect requiring years of training of Khmer rice scientists, at a minimum a centrally-located air-conditioned dehumidified room in Phnom Penh is needed to prevent deterioration of important international and traditional varieties. Due to lack of 24-hour electricity, cold storage at the provincial level is not practical (except perhaps in one or two locations using generators). At the village and farm levels, traditional methods of storage from harvest to planting are apparently adequate, although improved designs using local materials might minimize losses due to insects, rats, and heat and water damage.

Given the shortage of trained Kampuchean rice scientists and agronomists, investment in expansion of indigenous research capacity will not bring immediate benefit to peasant farmers. This very shortage, however, does have a positive side to the extent that farmers become part of the research process. Both in Kandal and Kompong Speu, in the area of the Prey Phdau Agricultural Station, farmers have participated in seed comparisons, fertilizer trials, and other experiments. Thus, aid to Prey Phdau, at present a simple thatch hut surrounded by rice fields on the grounds of the 1960s UNDP project, would not only bolster the Agronomy Department's ability to conduct field research but would also, in effect, support the training of more farmers in the use of modern seeds. The Tuol Samrong Research Station in Battambang is already receiving Soviet aid in the form of material and personnel.

Fertilizers

Contrary to the myth of Kampuchea as a "fertile land," poor soils predominate away from the river bank land enriched by nutrients deposited during the annual flooding of the Mekong basin. Despite this general lack of fertility, Khmer peasants have not traditionally worked to improve the soil through application of animal manure or nitrogenous plants. This involves intensive, back-breaking labor and, as noted above, labor in Kampuchea is scarce. During the period of mass mobilization under the Khmer Rouge, cadre organized the application of biofertilizers, including human wastes and even, survivors maintain, human corpses. These practices, even the sensible ones, have had no noticeable carryover in the post-Pol Pot period. Kampuchea thus lags well behind its Asian neighbors in the application of biofertilizers on rice soils.

Several of the voluntary agencies have attempted to interest the national Agronomy Department in experimenting with blue-green algae, a nitrogen-fixing organism which multiplies rapidly in specially-built tanks and then must be plowed into moist paddy soil. OXFAM staff hand-delivered some Burmese innoculant of blue-green algae to the Agronomy Department, but subsequent attempts to propagate it failed. This failure reflects the lack of interest at the national level in alternatives to chemical fertilizers. The entire orientation of the Ministry of Agriculture is towards encouraging the import of chemical fertilizers on a grant basis. This policy is very short-sighted, but understandable in the context of the relief effort which placed unprecedented amounts of chemical fertilizers into the hands of Kampuchean rice farmers. While the last Sihanouk five-year plan envisaged construction of a nitrogen fertilizer factory in Kompong Som, such grandiose dreams are well beyond the means of Kampuchea today. As aid from the international organizations winds down, the scarcity of chemical fertilizers, which produce no lasting benefits for nutrient-poor soils, will retard the expansion of the hectarage of fertilizer-dependent high-yielding rice varieties.

What rice farmers have traditionally relied upon, however, and now miss greatly as demand far outpaces the revival of production, is locally-crushed ground rock phosphate. Phosphate deposits exist in Kampot and Battambang provinces. Factories to exploit this resource existed near both deposits, with the Chinese-built plant in Kampot, sabotaged by retreating Pol Pot troops in early 1979, having a capacity of about 12,000 mt per year.[83] Kampuchea's soils are phosphate poor, particularly in the eastern provinces of Prey Veng and Svay Rieng, so this amount met only a portion of the country's needs. Illegal trade in super-phosphate fertilizer, available in abundance in South Vietnam during the war, also provided Khmer peasants with this nutrient.

Only production at the smaller plant in Battambang has revived since 1979. Production in the first half of 1983 was about 1,200 mt, far below the needs of the western provinces, much less the entire country.[84] Peasants bring their ox carts as far as 100 kilometers to purchase this valuable commodity at the factory site, particularly in April when seeding begins in floating rice areas. While the rock crusher itself was repaired with assistance from Oxfam America, lack of transport for the rock to the factory and inefficiencies in the secondary grinding after the initial crushing limit further expansion in production. Oxfam America's attempt to supply needed transport to the factory was initially blocked by the State Department, but finally in early January 1984 the denial was reversed on appeal. But further aid is needed to develop this and other alternatives to imported chemical fertilizers. In Vietnam and India, however, Kampuchea has allies who are doing some of the best work in the world in this area. Yet the transfer of appropriate, self-reliant, but labor intensive technologies to Kampuchean peasants will take many years, retarding the improvement of rice yields per hectare which remain among the lowest in the world.

Irrigation

The Pol Pot government made solving Kampuchea's water problems its number one development priority. In the absence of irrigation, much of Kampuchea's rice land lies fallow in the dry season from December to April. During the monsoon season, irregular rains plague peasants with droughts and floods, depending on the time of year and the maturity of the rice plant when the rains finally come. To liberate Khmer peasants from the vagaries of the weather, the Khmer Rouge mobilized thousands of people to build the long irrigation canals and large dams which today remain as memorials to the men and women sacrificed to their construction. These canals hold rain and stream water in the monsoon season but are often as bone-dry as the surrounding fields during the dry season. While some of the Khmer Rouge period projects are worthless, others were soundly designed and could provide a basis for further development of Kampuchea's irrigation infrastructure.[85]

With North Korean material and technical asistance, the Khmer Rouge constructed seven large pumping stations capable of irrigating up to 10,000 ha each. The engines powering the pumps, however, require extensive repairs; in addition, the fuel demands of the stations are quite large, belying once again the notion that the Khmer Rouge relied exclusively on primitive or (more positively) appropriate technologies. The large gravity-fed systems built during the Pol Pot period suffer from inadequate maintenance of dikes and channels. Kampuchean teams trained on heavy earth-moving equipment by Cuban hydrologists employed by Church World Service have performed major repair jobs in the past several years, notably saving 20,000 ha from annual flooding due to faulty design of a Khmer Rouge project in Prey Veng. These teams have also constructed several new earthwork systems themselves. On a daily basis, however, peasants have not been mobilized to shore up channel walls and keep them free of weeds and silt. As a result, water collects in the channels of these systems, but the systems themselves do not function as intended. This forces peasants to use traditional pumps to lift the water into their fields or to resort to growing rice in the channels themselves.

Thus, one major irrigation need in Kampuchea is an efficient, labor-saving pump which would enable peasants to utilize water available in ponds, streams, and man-made channels during the rainy season. Small diesel-powered pumps have been useful for keeping seed beds watered or getting enough water into a paddy for plowing. But maintenance and fuel problems have plagued these pumps continually. In the dry season, when small diesel pumps would be most useful, peasants cannot depend on enough fuel being available for an entire season to risk putting labor into a crop which will not be irrigated through maturity. For small tasks, traditional pump designs such as the pedal-powered *rohat* and the *snach*, a basket suspended from a wooden tripod, are adequate and reliable, but require large amounts of labor. Oxfam America was able to obtain a license to send tools to repair *rohats*.

Modern human-powered as well as wind- and solar-powered pumps, should be tried in Kampuchea. Prototypes of promising small pump designs would provide the Department of Hydrology with the opportunity to reach its own conclusions about alternatives to traditional designs. Oxfam America's application to send 10 solar-powered irrigation pumps to Kampuchea was denied by the State Department. Further, like their counterparts in the Agronomy Department, the technicians in Hydrology were spoiled during the emergency period by the abundance of imported pumps and fuel to power them. Now they face the necessity of repairing existing systems, mobilizing farmers to maintain dikes and canals, gradually expanding their own capacity to construct new water management schemes, and adopting alternative pump designs for trial and replication. With the exception of mobilizing farmers, primarily a political task, these other challenges can only be met successfully with outside assistance.

Training

This is one area crucial to Kampuchea's agricultural recovery for which the PRK has not sought significant amounts of aid from the international organizations. Even during the emergency period agencies were lining up to provide material support for agricultural training institutions in the area near Phnom Penh and foreign experts to train cadre in a variety of fields. The urgency of training was obvious given the severe shortage of trained personnel in agriculture, another legacy of the Pol Pot period. Not only were many killed from 1975 to 1979, but many survivors chose to go into exile in the early days of the Heng Samrin government. In 1980 there were probably more trained Khmer agronomists in Paris than in Kompong Cham, where only *four* technicians were responsible for agricultural development in this province of one million people.

The PRK prevented international agencies from participating directly in training activities because technical training includes a healthy dose of political training to orient new cadre towards socialism and solidarity with Vietnam. When the private agencies have participated in training, it has largely been in informal settings where learning by watching and imitating is the rule. Examples of this type of training include instruction in the use of earth-moving equipment or in animal vaccination techniques. Material aid has included school supplies, vehicles, and building materials for Chamcar Daung and Prek Leap Agricultural Colleges. Oxfam America provided school supplies and laboratory equipment to Chamcar Daung in early 1981.

Agricultural training abroad is taking place primarily in Vietnam. Training abroad has removed numerous talented young people and even experienced technicians from the country for extended periods, up to five years in some cases. The Heng Samrin government has been willing to give up the contributions these people could make to present reconstruction efforts in Kampuchea for the skills and political attitudes which the trainees will bring back with them from abroad and apply to solving the country's long-term problems.

Inside Kampuchea, the Ministry of Agriculture has shown a refreshing pragmatism thus far. At Chamcar Daung Agricultural College, courses have been limited to three-month intensive training sessions in tractor repair, water management, veterinary science, and other subjects. Students are nominated and sent by the provincial administration; upon completion of the course they are expected to return immediately to the provinces and get back to work. While the effectiveness of the crash courses has not been studied, their practical work orientation is laudable.

Certain tensions within the government have a significant impact on training designs. Many government officials studied in Europe, the United States, or Kampuchean institutions patterned after the French system. What they envisage as appropriate models for training differs radically from that of a long-time communist cadre with little formal education, or education in Vietnamese and Soviet institutions.

In agriculture, these conflicting approaches will have to work themselves out at the Agricultural College from which a new generation of agricultural technicians will emerge over the next decade.

While the present bent of the courses is decidedly practical, the one-time Director of the College, now a mid-level official in the Ministry, produced the 1968 catalogue of the Royal Agricultural College to give visitors an idea of what type of curriculum the College intends to offer.[86] Whether such an elite model for education can be tolerated in revolutionary Kampuchea remains to be seen; however, it is probable that one reason for the delays in the start-up of the two-year degree-granting institution has been lack of consensus regarding the return to the methods of the 1960s. Practical considerations—lack of books, lab equipment, dormitories, etc.—have also played a part in delaying the re-opening.

Vietnam and India have crucial roles to play in training people to manage Kampuchea's agricultural development. Both are rice-producing countries which must obtain maximum yields at every planting to feed their populations. Both have experience with high-yielding varieties, indigenous seed development programs, and alternative technologies. The ability of their specialists and training institutes to communicate their knowledge effectively to their Khmer counterparts will be a crucial determinant of Kampuchea's agricultural development over the next decade.

AID RESPONSES

Private Voluntary Organizations

Oxfam America has focused its post-emergency program on as many of these crucial obstacles to agricultural recovery as its limited budget would allow. It has introduced internatonal rice seed varieties better suited to Kampuchean conditions in the rainy season than IR 36. It has facilitated the transfer, noted above, of 140 traditional Khmer rice varieties from the Germplasm Bank of IRRI to the Ministry of Agriculture. Oxfam America's most important contribution has been to invest more than $150,000 in the Battambang Phosphate Factory, now repaired and supplying local peasants with ground rock phosphate fertilizer. With NOVIB, a Dutch agency, Oxfam America is funding the repair of two large irrigation stations built during the Pol Pot period at Ksaich Sar and Po Leus in Prey Veng province.

Other American agencies have also continued to make important contributions to Kampuchea's agricultural recovery. AFSC and CWS have assumed responsibility for supporting the Department of Veterinary Medicine's country-wide animal vaccination program. CWS currently has assigned an expert to the Department of Agronomy in Phnom Penh to help develop a storage program for vegetable and rice seeds. It has also contributed irrigation specialists who have trained a team of Khmer technicians to operate earth-moving equipment urgently needed to repair damage to large irrigation dams and dikes caused by neglect or faulty design. MCC has supplied 1,000 tons of food to the Ministry of Agriculture to be used on food-for-work projects related to repair and maintenance of irrigation systems. This rice will be used to feed local workers repairing the canal systems of the two large irrigation stations in Prey Veng.

American private agencies, therefore, along with European counterparts such as OXFAM, NOVIB, and CIDSE, have had to assume a role in Kampuchea's reconstruction normally filled by U.N. agencies such as FAO and UNDP. For the U.N. Kampuchea resolution expressly forbids the U.N. agencies from providing crucial inputs for long-term reconstruction and development. Private humanitarian agencies, which in countries like Thailand, the Philippines or even Vietnam normally fill the gaps in villages neglected by massive multi- and bi-lateral aid programs, have had their resources and technical capacity stretched to the limit by the Kampuchea program. Further, the resources at the disposal of private agencies during the emergency period created the expectation among the officials of the Heng Samrin government that OXFAM, for example, would continue to supply spare parts and raw materials to the textile factories in Phnom Penh indefinitely. Few of the agencies still working in Kampuchea have anywhere near the resources now to maintain their former roles as supplier of Western spare parts and raw materials. But in the context of the economic blockade against Kampuchea by Asian and Western capitalist countries, it is precisely these inputs which are most urgently needed. This fundamental dilemma cannot be resolved by the private humanitarian agencies which have budgets for Kampuchea which limit them to the small-scale provision of key materials which will have a long-term impact.

Oxfam America cannot become a mini-FAO; nor can any other agency fulfill this role. Kampuchea, like other developing countries, requires assistance from the U.N. system. Yet now even American private agencies find their limited aid programs in the agricultural sector blocked by U.S. government policy which applies the same criteria to the agencies as they have applied as major donors and policy makers to the U.N. program. By prohibiting American agencies from providing tractor spare parts, seed storage equipment, solar irrigation pumps, school supplies and training manuals, animal vaccines, etc. this policy precludes entirely the possibility that American humanitarian organizations would be able to substitute adequately for U.N. programs already effectively constrained.

Soviet and Vietnamese Aid

The U.N. prohibition on development assistance stems in part from a frustration with the Soviet and Vietnamese roles in Kampuchea. Neither country participated in the emergency program mounted by the United Nations. There is a sense among major donor countries, not only the United States, that the Heng Samrin government's major allies got off rather cheaply during the famine and period of immediate recovery; that is, the Western capitalist countries saved the People's Republic of Kampuchea for domination by the Soviet-led communist bloc. Donor countries argue that they have given enough; it is time for the Soviets and the Vietnamese to begin footing the bill for reconstructing and developing the client state of their creation.

That the Soviets and the Vietnamese treat the content of their aid programs as classified information only fosters the view of the U.N. donors that non-military bilateral aid is negligible. Quite early on in the relief effort, in November, 1979, the Soviet Ambassador to the United Nations, Vladimir V. Shustov, wrote a letter to the *New York Times* to counter charges that the Soviet Union was doing nothing to help the beleaguered people of Kampuchea. At that time he listed the components of an $85 million emergency program: 159,000 tons of cereals, $1.5 million worth of medicines, 660 transport vehicles, 50,000 tons of petroleum products, 7,800 tons of rolled iron articles, 5,000 tons of cement, 903 tons of paper, and 4.6 million meters of cotton and silk fabrics.[87] This is a rare detailed accounting of Soviet assistance. The 1982 FAO Report states that aid "granted by one of the most important bilateral donors" (presumably the Soviet Union) amounted to $300 million since the end of the emergency period, although another estimate is significantly lower, about $160 million, for 1981-82.[88] According to FAO, bilateral assistance was related to air and road transport, power, fuel, and fertilizer and major efforts are underway for more development-oriented activities in the fields of agriculture, health, and training.

Aid to Kampuchea's battered transportation sector is probably the most visible of the large bilateral aid projects. Hundreds of Soviet and East German trucks, buses, and passenger vehicles form the backbone of Kampuchea's internal road transport network. Since the end of the emergency period the Soviet Union has supplied the bulk of the fuel to power not only these vehicles but industrial machinery as well. Several Soviet passenger jets constitute the small fleet of Air Kampuchea. The Soviets have contributed equipment and dock workers to increase the efficiency of the deepwater port at Kompong Som. An agreement signed with the PRK in June 1982 committed the Soviets to providing another 50 trucks, earth-moving equipment and tar for road building, as well as 200 tractors and 10,000 mt of urea fertilizer to increase agricultural production.[89]

The Soviets have not fulfilled several important commitments, however. For several years, despite interest from the voluntary agencies, the repair of the buildings and classrooms at Chamcar Daung Agricultural College outside Phnom Penh has been reserved for the Soviets. To date the College still has not opened, not only because of conflicts over the curriculum design alluded to above, but also because of delays in the repairs needed for the physical plant. Another major Soviet project is the repair of the Chinese-built phosphate factory at Tuk Meas in Kampot. Despite numerous assessment missions by Soviet experts, no work has begun on this extremely important facility. Reopening was originally scheduled for 1984, but it is clear that this timetable cannot be met. In contrast, the Phnom Thom phosphate factory in Battambang, repaired with assistance from Oxfam America and CIDSE, a consortium of European Catholic agencies, began production in 1982 and has been working steadily (but below capacity in 1983). Apparently, repair of the Tuk Meas plant does not involve a huge investment; Ministry of Industry officials are somewhat mystified by the Soviet delays. These officials are considering re-designating the factory a voluntary agency project to get the repair work underway as soon as possible. To re-assign responsibility for a project involves a delicate process of political negotiation which is not always successful, as in the case of the Agricultural College where Kampuchean frustrations have been building for many months without noticeable results.

Further, the Soviets are no more likely to provide appropriate development assistance to a small, terribly underdeveloped country like Kampuchea than any other industrial giant. This places the Kampucheans in the position of accepting Soviet assistance which inevitably involves long-term dependence on a type and scale of technology which may be inappropriate. For example, the Soviets are providing generators for electrification projects in the provincial capitals where the supply of electricity has been fitful since 1979. In Battambang town, however, rather than providing two or three medium-size generators, the Soviets have donated a single large one which weighs about 36 tons. The Kampucheans did manage to get the generator on a train in Phnom Penh and get it up to Battambang. The generator remains at the railway station for the time being because the Industry Office in Battambang has no way to transport it from the station to the site. On the site itself, four Soviet technicians and a Khmer counterpart who studied Russian in the 1960s have to design and construct a completely new building to house the massive generator. The wood for the project is lying by the side of the road in a small town in Pursat province for lack of transport to Battambang. When the generator is finally installed it will consume about 8,000 liters of fuel, or one tank truck, per day in a country already saddled with serious fuel shortages.

Vietnamese aid, of course, cannot be on such an inappropriate scale as Vietnam itself has food and technical problems almost as daunting as those of Kampuchea. From the beginning of the emergency period, Vietnam's major contribution has been the provision of experts in Kampuchea itself and the training of Khmer cadre in Kampuchea and in Vietnam. As noted above, this aid is the key component of the Vietnamese effort to create a competent PRK administration which is loyal to Vietnam. Vietnam has provided material assistance, of course, including food and medical aid in 1979 during a period when Vietnam was experiencing severe economic problems including serious food shortfalls. Vietnam has also instituted a system of twinned provinces as a way of providing assistance. A Vietnamese province is linked to a Kampuchean counterpart; aid goes from the Vietnamese province to the Kampuchean one, while visits and expressions of solidarity are exchanged regularly. This type of aid is difficult to quantify, but Phnom Penh radio regularly cites examples of province-to-province aid including, in one instance, 60 tons of rice seed and help in constructing dispensaries, a radio station, a sawmill, a veterinary post, and two reservoirs. Overall, the Vietnamese friendship treaty with the PRK commits them to providing at least $25 million in annual assistance.[90]

India sent an assessment team in January 1982 to look into the possibility of providing bilateral assistance. At the time this visit seemed to hold considerable promise, for not only would Indian aid bring welcome diversity to bilateral aid otherwise dominated by the Soviet Union and its allies, but India is a world leader in the application of appropriate, indigenous technology to rice production and village-level industries. India has failed to fulfill the expectations raised by this visit. While some Khmer technicians are being trained in India, actual material aid has been negligible, lower than even several voluntary agency programs. Apart from a modest commitment ($1 million) to advancing small industries such as edible oil production, nothing has resulted from the assessment mission.[91]

This failure, coupled with the Western blockade on development assistance through the U.N. system, means that for the foreseeable future the Heng Samrin government will be totally dependent on the Soviet bloc. This is not an enviable position to be in for a country with so many urgent needs. Government officials with training in capitalist countries during the Sihanouk or Lon Nol regimes make no secret of their frustration with this state of affairs. "Would you rather be the servant of a rich master or a poor one?", they ask pointedly. Kampuchea must be dependent on outside aid for many years, but some officals wonder why they are forced to be dependent on countries with serious economic difficulties of their own.

PRK Government Policies and Reconstruction

In this context, effective government policies which husband scarce resources are essential. Yet seldom has a country been less ready or suited for self-reliance than Kampuchea in the 1980s. The government is composed of cadre from disparate backgrounds with little experience in governing a socialist country. Further, an important legacy of the Khmer Rouge period has been to thoroughly discredit self-reliant approaches to development which depend on mass participation. Under Pol Pot mass mobilization to dig irrigation channels, clear new land for rice production, and exploit timber reserves involved minimal use of fossil fuels and virtually no imported resources. But the human cost of this mass mobilization was simply too great. Now efforts to organize work teams to fertilize rice fields with biofertilizers or to maintain irrigation works constructed under the Khmer Rouge fail amid fears that this is but the first step towards a reversion to the methods of Pol Pot cadre. Thus, the task of organizing a self-reliant response to Kampuchea's current problems is made virtually impossible by the perception that socialism will inevitably lead to the return of the excesses of the Khmer Rouge.

This fear of socialism, or at best cynicism about what it can accomplish in Kampuchea, undermines the allegiance of many "intellectuals," those who received a Western education in pre-1975 Kampuchea, to the Heng Samrin government. In turn, the committed communist leadership of the PRK suspects these cadre of secret sympathy with the non-communist resistance. These suspicions reinforce each other and undermine the stability and effectiveness of the government. The problem is that trained cadre are in such short supply that a regime in Kampuchea unable to attract at least modest allegiance from the intellectuals will not be able to oversee the recovery. The best example of the strength of a pragmatic approach is Battambang where the bulk of the provincial administration is composed of intellectuals whose technical judgement is respected. They chafe at some of the political control, but in the words of one technician, "We are allowed to do our job." The remarkable economic revival of Battambang province owes a great deal to its proximity to the Thai border and to its abundance of rice land. But the rehabilitation of Battambang's industrial plant, including key industries such as the jute sack factory and the phosphate mill, and the overall organization of the recovery, result directly from a willingness to utilize the talents of skilled people, even if they do have diplomas granted in France or California. Elsewhere in Kampuchea intellectuals are under much greater political pressure because of their backgrounds. This pressure makes many depressed and ineffective. It drives others into exile.

In practice, however, the Heng Samrin government in many ways has been socialist in name only—internationally, but less so in its actual domestic economic policies. In the countryside, peasants are organized into *krom samaki* (literally "solidarity groups") which function as mutual assistance teams at peak work periods, primarily during plowing, transplanting, and harvesting the rice crop. But peasants in these groups retain private ownership of the means of production. While land is in principle owned collectively, in land-rich areas such as Battambang peasants speak openly of private plots and clearly farm certain areas as individual families. In theory, the harvest on collective land is divided among solidarity group members according to work points, including contribution of means of production such as draft animals, and need. With so many women heading families in Kampuchea today, the primary advantage of this loose cooperative organization is that it provides for the labor and subsequent food needs of these families. Without some form of mutual assistance they would be hard pressed to care for themselves.

After the distribution of the harvest, the rice crop becomes the private property of the family to store or sell as their needs dictate. Remarkably, the government has yet to impose taxes on the peasantry. The PRK authorities rely completely on the powers of persuasion and whatever cash or consumer goods they can muster to get the peasants to part with a portion of the crop. Thus, just before the harvest, particularly in surplus areas such as Battambang, cadre travel from village to village emphasizing the patriotic duty of the peasants to sell rice to the government so that stocks can be built up and town dwellers fed. (A popular song which reinforces this message is entitled "Selling Rice to the State.") Naturally, like their brethren all over Southeast Asia, Kampuchean peasants were disappointed to be offered only 1.4 riels (about 3 cents at the free market exchange rate) per kilogram of paddy after the 1982 harvest. Despite this low price, and the paucity of consumer goods offered by the government for barter, the PRK authorities were able to purchase 160,000 mt of paddy from surplus areas after the 1982 harvest.[92] This probably reflects not only the cash needs of the Khmer peasantry for investment in new oxcarts, draft animals, houses, and consumer goods, but also the surprising degree of political integration of the peasantry with the Heng Samrin government.

As impressive as the 160,000 mt figure may be in the context of the disastrous food situation of the previous decade, this still does not represent the achievement of self-sufficiency for Kampuchea. Not all of this rice is surplus in the sense that families which sold it had adequate stocks for all of 1983. Many families sold rice in anticipation of dry season or early rainy season harvests to come. Drought conditions in 1983 meant serious hardship for families who made this choice. Furthermore, the food needs of Phnom Penh alone amount to 4,500 mt of rice per month, the equivalent of about 90,000 mt of paddy annually, 56 percent of the total purchased by the government in 1982-83.[93] In years in which production drops due to weather problems, Phnom Penh will demand virtually all of the country's surplus, leaving little food for distribution in rural deficit areas.

The weakness and poverty of the Heng Samrin government leaves Khmer peasants vulnerable to the re-establishment of inequitable creditor-debtor relationships in the countryside. Traditionally, the village merchant, either Chinese or Sino-Khmer, mediated virtually all of the Khmer peasantry's commercial relations with the outside world.[94] Goods purchased from the local shop during the planting season were paid for with paddy at harvest times, at interest rates of 100 percent. The merchant milled the paddy and then sold it to the state or to rice merchants with trade networks extending to Saigon, Hong Kong, and Singapore. This export market has not yet been established. And these village merchants were an obvious target of oppression under the Pol Pot government. But the very existence of private rice mills side-by-side with state mills charging fixed government prices as much as 75 percent lower for their services suggests that a system of private control of at least a portion of the rice market may have returned.[95] That the government lacks the means to loan resources to peasants during the "hungry period" before the harvest would also suggest that moneylenders may be filling this void at the village level.

In the towns the government share in the markets is extremely low. The authorities collect taxes from vendors but otherwise tolerate the free market. During the emergency period goods poured into Kampuchea from Thailand in exchange for gold and dollars which had been hidden away in secret caches during the Pol Pot period. Tensions at the border and the imposition of customs duties on imports from Thailand have drastically reduced the flow of consumer goods, although Thai textiles continue to be imported via ocean transport, arriving in Kampuchea officially through an island off the southern coast. Vietnamese consumer goods also claim a significant portion of the private market. But prices in these markets are quite high and turnover is low, even for the impressive array of local foodstuffs available.[96] Although they receive food and clothing supplements, government workers in Phnom Penh have little cash with which to purchase goods in the free market. Their salaries are so low that most families require spouses, siblings, and children to engage in supplemental income-generating activities, such as maintaining a market stall or becoming a pedicab driver. Some government workers engage in similar activities themselves. The danger is that Phnom Penh will begin to resemble Ho Chi Minh City where the worst cases of malnutrition are found among the children of government cadre unable to feed their families on their pitifully low salaries.

Low salaries drive people to the private sector. A skilled mechanic, for example, can earn a standard government salary of 100 or 120 riels per month working in the garage of a Ministry in Phnom Penh or can do the same work in a private repair shop for four to five times this amount. The urban service sector—restaurants, pedicabs, tailor shops, repair shops—is almost entirely in private hands. Government revenues from taxes on this sector amount to only 1 million riels ($25,000) out of a total budget for the city of 40 million riels ($1 million).[97] As prices are significantly higher in the private sector, the temptation to engage in small-scale corruption—such as selling products of state enterprises on the private market—is great for poorly-paid government workers. Truck drivers are in a particularly strong position because they have access to fuel. A Ministry of Industry official described how drivers deliberately break their odometers, get permission for a long cross-country trip, receive their fuel allotment, shorten their trip, and finally sell the excess fuel on the free market. To compound the bonanza, drivers carry not only cargo but passengers who pay for rural and cross-country transport service. While no one can build villas with the proceeds, as in the Sihanouk and Lon Nol periods, petty corruption of this sort slows Kampuchea's recovery. The state has so few resources that to use them inefficiently is disastrous. The case of fuel is particularly serious, for as Hondas and trucks filled with people ply the roads, diesel irrigation pumps and productive enterprises such as the phosphate factory stand idle for lack of fuel.

Even a city as poor as Phnom Penh attracts people from the countryside looking for opportunities to participate in the trade or service economy and gain income for their families. These people come especially during seasons of low employment in the rural areas. The head of the Revolutionary Committee of Phnom Penh, Keo Chanda, has discussed plans to stem this tide of unemployed people by moving them into green zones around the city where food will be grown on state farms or cooperatives to supply the city's population. Phnom Penh is now a city of 600,000 which is tiny by Asian standards; but it is a city which completely lacks urban services. Indeed, most of Phnom Penh's residents lived in rural areas before 1970 and there is a village atmosphere to the city which unfortunately translates into poor sanitation, chaotic settlement patterns, and what might charitably be called a cavalier attitude towards maintaining water and electrical systems. The green zone scheme would not only help ease some of the crowding, but would be a modest effort towards giving the state a share in the marketing of food in the capital. Whether movement of people to these zones would remind people too much of Pol Pot remains to be seen; in any case, the comparison was not lost on at least one Western journalist who interviewed Mr. Chanda in March 1983.[98]

On a national level, the PRK lacks the means to generate foreign exchange except through bilateral aid programs. The prospects for resumption of rice exports are poor given the continued lack of progress in expanding the amount of cultivated land coupled with the 1983 drought and floods. Kampuchea did export a small amount of corn in 1982 to Singapore.[99] Singapore merchants are anxious to resume trade with Kampuchea for freshwater fish and shellfish from the Tonle Sap Lake and the Mekong basin. For the time being, however, the PRK has agreed to limit exports to 15,000 tons annually to Vietnam.[100] Rubber production for export has been limited by the legacy of the destruction of the war and the lack of investment in new trees and maintenance of processing facilities over the past 10 or 15 years. East Germany has agreed to loan the PRK $10 million for reviving production of this important export commodity. Virtually all of the 8,000 tons currently being produced goes to the socialist countries, primarily the Soviet Union.[101]

There are a few glimmers of hope that Kampuchea will be able to resume trade with its capitalist neighbors. The model for this trade is Vietnam's barter arrangements through state import-export firms in Ho Chi Minh City with Singapore-based trading companies. The Vietnamese export primary products and receive manufactured goods, primarily spare parts, for production facilities established by private capital or U.S. aid before 1975. Kampuchea will also require key inputs from the West which can no longer be supplied in sufficient quantity by the international organizations. Singapore merchants who conduct the barter trade with Vietnam are anxious to begin similar arrangements with the PRK. In the west there may be clandestine contacts between Battambang provincial officials and Thai capitalists regarding the export of timber and other forest products. Samples of possible exports, including spices and rare medicinal plants, have been received by a Bangkok-based trading firm. There are even rumors in Phnom Penh that PRK officials have discussed trade directly with Thai counterparts at Don Muang International Airport in Bangkok. These possibilities, while promising, have yet to come to fruition and will probably be of marginal utility as long as the political atmosphere between Indochina and ASEAN remains polarized.

There is an anarchic quality to life under the Heng Samrin government which belies its draconian image to the outside world. The people of Kampuchea do not pay taxes; they do not pay for electricity and water in urban areas; they have refused to participate in rural organizing efforts; they profited from illicit trade across the border controlled by the government's principle enemies. The Heng Samrin regime lacks the resources and the skills to organize effectively long-term efforts towards economic independence and self-reliance. While some Kampuchean government policies have contributed to the lack of progress towards developing self-reliance, the primary obstacles to development in the post-emergency period have been the 1970s legacy of human and physical destruction and the cut-off of U.S., U.N. and other western assistance crucial to Kampuchea's recovery.

FUTURE PROSPECTS

The people of Kampuchea entered 1983 saddled with a food deficit of 100,000 mt.[102] In making these calculations, FAO uses the World Health Organization figure for minimum rice rations of 12 kg per month. This figure is a *minimum*, not an optimal amount for men, women, and children facing the immense reconstruction tasks of their country. FAO makes calculations based on aggregate production figures, which are insensitive to differences in production capacity and distribution in surplus provinces versus deficit ones. If in July a family in Battambang has one ton of paddy stored beneath its house, then another family in Prey Veng or Takeo has exhausted its reserves and has no stocks remaining until the next harvest.

These food production shortfalls have enormous human costs. According to a nutritional survey conducted by the FAO, in 1982 53 percent of the children under 12 suffer from moderate to severe malnutrition.[103] The prevalence of malnutrition rises sharply after children have been weaned: 36 percent for children up to one year, 62 percent for children one to three years, 59 percent for children four to six years, and 62 percent for children above six years of age. In addition, a majority of older children have suffered severe growth retardation. "Children with ages ranging between 8-12 years look like children of 5-8 years. This severe under-development, or nutritional dwarfism, is the result of prolonged malnutrition during the first few years of life."[104] The eight to twelve-year old children were born in the early 1970s under the Lon Nol regime and are therefore part of the "lost generation of children" predicted by a World Vision doctor in 1975.[105] FAO concludes, "At the present time, not enough food is available in Kampuchea, especially at the village level, to meet even minimum acceptable level [sic] of adequate food intake. It is no longer desirable to take the risk of more damage being caused to the young generation of Kampuchean children, who will be the future citizens of this country."[106]

This statement would hold true for many developing countries. What makes Kampuchea unique is that even as nations acknowledge its suffering they refuse to send development assistance on political grounds. Thus, in 1983 Kampuchea received virtually no aid from the U.N. system which as a matter of course provides millions of dollars to other nations suffering from problems similar to those of Kampuchea. Many of the problems of Kampuchea are man-made, created in part by the very governments, including that of the United States, that now deny aid to the Khmer people.

The last FAO report, quoted above, was written in anticipation of an excellent 1982 monsoon season rice harvest. In 1983, however, Kampuchean peasants fell victim to the unusual weather patterns which prevailed in many parts of the world. In the eastern provinces, drought was severe early in the planting season. In Prey Veng, for example, drought prevented farmers from planting the planned 23,000 ha of short-season rice, normally planted with the first rains and harvested in August. Only one-third of the planned hectarage for vegetables was planted. In Kompong Speu, as of the beginning of August, it had rained only seven times since April, when the rainy season begins. Out of a projected 75,000 ha for rice cultivation, only 1,150 had been transplanted. The normal figure for late July-early August is 35,000-45,000 ha. In this situation, flooding became the problem in September when the rains finally did come, drowning the small seedlings. The drought struck hardest in areas of Kampuchea which already show food deficits and are among the poorest in the country even in normal times.

In the west, Battambang, the rice bowl province, appeared in July to have enough rain for a successful harvest after early drought limited the hectarage of floating rice areas. In October, however, in a disaster typical of the vagaries of the monsoon which Khmer peasants are helpless to prevent or control, a typhoon which struck the coast of Vietnam penetrated far inland, causing flash flooding in northwest Kampuchea. The pressure of the water was so great that a 38 kilometer dike broke, flooding more than 150,000 hectares of Kampuchea's best rice land. A second dike had to be deliberately broken to save it from complete destruction. The flooding drowned oxen and domestic animals and washed away homes and food stocks. In the aftermath of the destruction, revised estimates of the

117

food deficit going into 1984 are close to 300,000 tons, the equivalent of the deficit at the end of 1980, only one year after the famine.[107]

This type of situation need not become a disaster. The government can intervene with small irrigation pumps to keep seed beds well watered or to pump water out of flooded fields. It can distribute rice seed and fertilizers to peasants who have lost their initial planting. Large irrigation systems can irrigate land which would otherwise be dry. Tractors can plow land where the soil is too hard for teams of oxen to break. People and equipment can be mobilized to begin the necessary repairs to damaged dikes. Surplus food can be distributed to needy families and the government can release stocks onto the market to keep prices down.

Because of the denial of development assistance, however, these means are not available to the Heng Samrin government. By refusing to aid with irrigation, by refusing to provide spare parts to maintain transport equipment, by refusing to help develop local fertilizer sources, by refusing to supply tractors and tractor spare parts, the donor countries to the U.N. program have prolonged the quiet crisis in Kampuchea and promoted disaster in years of drought or flooding. To provide food assistance *after* exacerbating conditions leading to food shortages is woefully inadequate from a humanitarian standpoint. More Khmer people have already joined Kampuchea's "lost generation" in the interim.

The weather problems coincided with an unsuccessful effort led by the Thai and Singaporean delegations to the United Nations to prohibit all assistance, even emergency food aid, to the Heng Samrin government. The 1982 U.N. Kampuchea resolution, "The situation in Kampuchea," adopted in October 1982, "*expresses its deep appreciation once again* to donor countries, the United Nations and its agencies and other national and international humanitarian organizations which have rendered relief assistance to the Kampuchean people, and appeals to them to continue existing arrangements to assist those Kampucheans who are still in need, *especially* along the Thai-Kampuchean border and in the holding centres in Thailand."[108] The Thais and the Singaporeans wanted all aid cut to the interior, which would have been neatly accomplished by deleting the word "*especially*" in the above clause relating to the U.N. aid program. While the United States did not object in principle to denying aid to the Khmer people inside the country, the State Department was concerned that such a ban would make the continuing border program look too political. The 1983 resolution, therefore, maintained the clause regarding humanitarian aid but added the term "emergency" to emphasize that no long-term assistance should go to the Heng Samrin government.

Even without an absolute aid ban, donor countries have simply refused to contribute to the interior programs in 1983. Sweden donated $1 million and the French offered 6,000 mt of wheat at a provisional pledging conference held in New York on September 16, 1983. In contrast, countries pledged more than $16 million to the border program, including $9 million worth of food rice from Japan through the World Food Program and $2 million cash from the United States to the U.N. Border Relief Operation that oversees the camps of the Khmer Rouge and of its coalition partners, supporters of Prince Sihanouk and Son Sann. Compounding its failure to contribute to the interior program in June 1983, the State Department denied three separate licenses to Oxfam America to ship 10 solar pumps for rice irrigation, transport equipment for the Battambang phosphate factory, and equipment to repair a large steam-powered rice mill.

The former decision was cruel given the drought conditions. AFSC has had a license renewal for vaccines and other veterinary supplies denied as well. In Europe, colleagues at NOVIB, a Dutch non-governmental organization, have informed Oxfam America staff of increasing Dutch restrictions on funding for Kampuchea. In Belgium, all requests for additional funding for Kampuchea projects of Belgian private agencies have been turned down except for a single project involving purchases on the Belgian market. The EEC has also refused recent requests for Kampuchea funding from Mani Tese, an Italian agency, and Oxfam Belgium. Thus, the refusal of the United Nations to send relief and development assistance to the interior is but one important component of a world-wide effort by Western capitalist countries to deny aid to the people of Kampuchea.

The crisis in Kampuchea is no longer the focus of world attention as it was in the fall of 1979. In a world of so much man-made suffering, government officials, the press, and ordinary people have difficulty maintaining interest in the long recovery and development phase after the most dramatic crisis has ended. What once was a dramatic, vivid, human situation becomes a matter for the usual cynical diplomacy of the East-West confrontation, mere fuel for propaganda machines.

The Khmer people have been victims for nearly 15 years—victims of bombing, invasion, civil war, revolution, famine. Now after a very brief period of recovery, the people of Kampuchea must again endure suffering caused by an economic blockade whose basis is the considerations of great power politics rather than the humanitarian needs of long-suffering people. Even in the best of circumstances, the Khmer people would need at least a generation to recover from the human disaster of the 1970s. By blocking humanitarian assistance, and leading the effort to ensure that other nations do the same, the

United States ensures that the trials of the Khmer people will continue indefinitely, and that another generation may be lost in Kampuchea.

1. William Shawcross, *Sideshow: Kissinger, Nixon and the Destruction of Cambodia* (New York: Simon and Schuster, 1979), especially Chapters 1, 8, 9, 11 and 19.

2. George Hilderand and Gareth Porter, *Cambodia: Starvation and Revolution* (New York and London: Monthly Review Press, 1976), p. 33.

3. Joel Charny, interview with Ministry of Agriculture spokesman, Phnom Penh, January 25, 1984.

4. James P. Grant, *The State of the World's Children 1984* (New York: UNICEF, 1984), p. 40.

5. David P. Chandler, *A History of Cambodia* (Boulder, Colorado: Westview Press, 1983), Chapter 3.

6. Ibid., p. 117.

7. Ibid., p. 151.

8. Ibid., p. 147.

9. Ibid., pp. 156-7.

10. Ibid., p. 162.

11. Shawcross, *Sideshow*, pp. 46-62.

12. Ibid., p. 64.

13. Malcolm Caldwell and Lek Tan, *Cambodia in the Southeast Asian War* (New York and London: Monthly Review Press, 1973), pp. 202-5, 228-237.

14. Shawcross, *Sideshow*, pp. 236-58.

15. Ibid., pp. 128-149.

16. One of the earliest reflections of this tendency is John Pilger's deliberately spectacular article, "Death of a Nation" in the London *Daily Mirror*, September 12, 1979. Most early OXFAM publications also attribute virtually all destruction in Kampuchea to Pol Pot and the Khmer Rouge.

17. Shawcross, *Sideshow*, pp. 264-272. See especially maps, pp. 266-7.

18. Kampuchean Inquiry Commission, *Kampuchea in the Seventies: Report of a Finnish Inquiry Commission* (Helsinki: Kampuchean Inquiry Commission, 1982), p. 12.

19. Hildebrand and Porter, *Cambodia: Starvation and Revolution*, p. 20.

20. Ibid., p. 20.

21. Food and Agriculture Organization of the United Nations (FAO), Office for Special Relief Operations (OSRO), *Kampuchea: Report of the Food and Agriculture Assessment Mission, October-November 1982* (Rome, December 1982), p. 20.

22. Hildebrand and Porter, *op. cit.*, p. 25.

23. Francois Pouchaud, *Cambodia Year Zero*, translated by Nancy Amphoux. (Harmondsworth, England: Penguin Books Ltd., 1978), p. 17.

24. Hildebrand and Porter, *Cambodia: Starvation and Revolution*, p. 25.

25. Ibid., p. 29.

26. Michael Vickery, *Cambodia 1975-1982* (Boston: South End Press, 1983), pp. 72-82.

27. Hildebrand and Porter, *Cambodia: Starvation and Revolution*, pp. 42-47.

28. Ponchaud, *Cambodia Year Zero*, p. 52.

29. Ibid., p. 115.

30. Michael Vickery's book, cited above, through careful analysis of numerous first hand accounts, demonstrates how varied conditions were in Kampuchea under the Khmer Rouge. Even in adjacent districts policies regarding the treatment of the former urban dwellers could be dramatically different. See Vickery's description of the Northwest Zone (Battambang, Pursat), pp. 100-20, especially noting the contrasting conditions in District 3 and Districts 2, 6, and 7, pp. 111-12, and 114-18. In the Eastern Zone (Prey Veng, Svay Rieng, eastern Kompong Cham) conditions were relatively good until the fear of Vietnam led the Pol Pot faction to destroy the Eastern Zone cadre, splitting the party irrevocably and leading to the Vietnamese invasion which ousted the Khmer Rouge. See, Vickery, pp. 131-38. It is beyond the scope of this study to describe the social and cultural policies of the Khmer Rouge — banning Buddhism, outlawing traditional forms of address, promoting communal living and forced marriages arranged by cadre — which virtually abolished traditional Khmer culture. These policies were as devastating as the Khmer Rouge development program.

31. Joel Charny has worked with a Kampuchean technician, now in the Heng Samrin administration, who survived the Khmer Rouge period because of the value of his tractor repair skills to the local cadre in Battambang.

32. Joel Charny, conversation with Massey-Ferguson representative in Singapore, March 1982.

33. Marie Alexandrine Martin, "La riziculture et la matrise de l'eau dans le Kampuchea democratique," *Etudes rurales*, juil.-sept. 1981, 37.

34. Ponchaud, *Cambodia Year Zero*, p. 103.

35. Kampuchean Inquiry Commission, p. 25.

36. Stephen R. Heder, "Kampuchea 1980: Anatomy of a Crisis," *Southeast Asia Chronicle*, 77 (February 1981), 4.

37. Martin, 11.

38. Vickery, *Cambodia 1975-1982*, p. 154.

39. Martin, 14.

40. Vickery, *Cambodia 1975-1982*, pp. 140-142.

41. Ibid., p. 144.

42. Michael Vickery, "Democratic Kampuchea — C.I.A. to the Rescue," *Bulletin of Concerned Asian Scholars*, Vol. 14, 4 (October-December 1982), 45-52, especially 51. See also Kampuchean Inquiry Commission, *Kampuchea*

in the Seventies, p. 35.

43. Ben Kiernan, "Vietnam and the Governments and People of Kampuchea," *Bulletin of Concerned Asian Scholars*, Vol. 14, 4 (October-December, 1979), 19-23.

44. Kampuchean Inquiry Commission, p. 36.

45. Conversation with Joel Charny, October 1981.

46. Jim Howard, "Brief details of visit by Jim Howard to Kampuchea and Vietnam, 24th August-7th September 1979," OXFAM, Oxford, England, September 9, 1979 (Typewritten).

47. Ibid.

48. Conversation with Joel Charny, October 1981.

49. Elizabeth Becker, "The Politics of Famine in Cambodia," *The Washington Post*, November 18, 1979.

50. Ibid. See also Kathleen Teltsch, "Cambodia Hobbling Foreign Relief Aid," *The New York Times*, September 23, 1979, and Henry Kamm's reporting from Bangkok in *The New York Times* throughout the fall of 1979, for example, "Hanoi is Said to Place Politics Before Lives of Cambodians," November 2, 1979.

51. The following account is based on Guy Stringer's colorful report, "A slow boat to Indo China, or the most expensive cruise in the world," submitted to OXFAM in Oxford, November 14, 1979. (Typewritten).

52. Linda Mason and Roger Brown, *Rice, Rivalry, and Politics: Managing Cambodian Relief* (Notre Dame, Indiana: University of Notre Dame Press, 1983), pp. 34-59.

53. Heder, "Kampuchea 1980: Anatomy of a Crisis," p. 10.

54. Mason and Brown, *Rice, Rivalry and Politics*, Chapter 3, pp. 91-134.

55. Mary McGrory, "Starvation, not regime, is enemy," *The Boston Globe*, December 17, 1979.

56. Rumor traced to U.S. Embassy in Bangkok by Oxfam America staff, July 1980.

57. National Foreign Assessment Center, "Kampuchea: A Demographic Catastrophe," (CIA, May 1980), pp. 2, 5, 6.

58. See William Shawcross, "The End of Cambodia," *New York Review of Books*, 21-22 (January 24, 1980).

59. Food and Agriculture Organization, Office for Special Relief Operations, *Kampuchea: Report of the FAO Food Assessment Mission*, W/P0180, (Rome, November 1980), Appendix C.

60. International Bank for Reconstruction and Development (World Bank), International Development Association, *Report of Economic Mission to Cambodia—1969*, (in three volumes), EAP-13a, (October 1970). Vol. I, pp. 13-14. (Hereinafter referred to as *IBRD*)

61. FAO Report, December 1982, ANNEX F, p. 92.

62. Conversation with Dr. Carmel Goldwater, Phnom Penh, October 1981.

63. Food and Agriculture Organization, Office for Special Relief Operations, Report of the Food and Agriculture Assessment Mission to Kampuchea, *23 October-4 November 1981*, W/P5662, (Rome, November 1981), p. 4.

64. FAO Report, December 1982, p. 44.

65. Ibid., p. 25.

66. International Committee of the Red Cross, *Kampuchea: Back from the Brink* (Geneva: n.d.), p. 37.

67. These and the following statistics on U.S. aid to the Joint Mission from ICRC, *Kampuchea: Back from the Brink*, p. 39.

68. Telex from Brian Walker to OXFAM, Oxford, October 1, 1979.

69. *IBRD*, October 1970, Vol. I, p. 83.

70. See Appendix, p. 137.

71. FAO Report, December 1982, ANNEX D, pp. 77-77b.

72. Ibid., ANNEX F, pp. 91-93.

73. Ministry of Agriculture estimates, October 24, 1983.

74. FAO Report, December 1982, p. i.

75. FAO Report, November 1981, p. 2.

76. Ibid., p. 45.

77. Jean Delvert, *La Paysan Cambodgien* (Paris and The Hague: Mouton and Co., 1961), pp. 235, 305, 322.

78. Ibid., pp. 242-243.

79. Interview with Meas Phanna, then Director of the Agricultural Machinery Enterprise, January 1982.

80. FAO Report, December 1982, p. 66.

81. Delvert, pp. 337-338.

82. Interview with Ministry of Agriculture officials, October, 1982.

83. *IBRD*, October 1970, Vol. I, p. 28.

84. Interview with factory manager, July 1983.

85. See Martin, pp. 23-37 for a detailed discussion of the irrigation program of the Khmer Rouge.

86. Interview with then Director of the Agricultural College, K.T. Bunthan, December, 1980.

87. Letters to the Editor, *The New York Times*, November 14, 1979.

88. FAO Report, December 1982, p. 35. For the lower estimate, see Sophie Quinn-Judge, "Kampuchea in 1982: Ploughing Towards Recovery," *Southeast Asian Affairs 1983* (Singapore: Institute of Southeast Asian Studies, 1983), p. 162.

89. Quinn-Judge, p. 162.

90. Ibid., p. 162.

91. Ibid., p. 162.

92. FAO Report, December 1982, p. i.

93. Ibid., p. 13.

94. Delvert, pp. 509-533.

95. This was acknowledged to Joel Charny by a mid-level official in the Heng Samrin administration, July 1983.

96. FAO Report, December 1982, pp. 15-16.

97. Interview with Keo Chanda, head of the People's Revolutionary Committee of Phnom Penh, March, 1983.

98. Colin Campbell, "Pol Pot's Bitter Legacy Weighs on Cambodia," *The New York Times*, April 5, 1983.

99. Quinn-Judge, p. 158.

100. Conversation with FAO consultant, October, 1982.

101. Quinn-Judge, p. 158.

102. FAO Report, December 1982, p. 49.

103. Ibid., p. 55.

104. Ibid., p. 55.

105. Hildebrand and Porter, *Cambodia: Starvation and Revolution*, p. 29.

106. FAO Report, December 1982, p. 56.

107. Information on drought was collected in Prey Veng and Kompong Speu, July, 1983. Flooding reports from Ministry of Agriculture conveyed by Eva Mysliwiec, OXFAM Field Director, Phnom Penh, October, 1983.

108. United Nations Assembly, Resolution 37/6, "The Situation in Kampuchea," October 28, 1982. Emphasis in original.

CHAPTER FOUR
Conclusion

For the Vietnamese and Kampuchean people the "Vietnam War" is not yet over. They live daily with the cumulative legacy of the war: defoliated forests and poisoned soils; destroyed schools, hospitals, and homes; loss of family members killed in battle or by revolutionary violence, starvation, and disease. In Kampuchea a guerrilla war continues with the civilian population caught in the middle of the conflict. War and its legacy have deepened the poverty of these peasant societies. Too many human and material resources have been devoted not to development but to destruction and makeshift reconstruction in a seemingly endless cycle.

The United States, employing the many weapons in its arsenal, has contributed inordinately to this destruction. The bombing and defoliation of vast areas of both countries represent the most massive use of conventional military force in history. With the military defeat of its allies in 1975, the United States adopted hostile policies towards the new governments of Vietnam and Kampuchea. Since 1975, the United States has waged diplomatic, political, and economic warfare to isolate the governments of Vietnam and Kampuchea. Far from seeking to heal the wounds of war the United States, with the exception of a brief period during the Carter Administration in 1977 and 1978, has sought to engage the Vietnamese and Kampuchean governments in wider geo-political conflicts.

The revolutionary governments of Vietnam and Kampuchea share responsibility for the plight of their people. In Vietnam, the leadership has failed to build a commitment in the south to their socialist vision of the country's direction. Economic policies emphasizing industry rather than agriculture proved inappropriate. Poor planning hampered the essential relocation of urban dwellers onto productive land. The international conflict with China and internal crackdowns on the Chinese-dominated free market drove ethnic Chinese families to flee Vietnam, depriving the country of skilled personnel. Others left because they lacked the revolutionary background required to play a substantial role in rebuilding the country.

In Kampuchea the Khmer Rouge sought the immediate destruction of the society created by colonialism and the relatively brief U.S. intervention. This destruction entailed not merely the emptying of urban areas, but the virtual enslavement of those who had been corrupted by values from the world outside the poor, isolated villages of rural Kampuchea. By putting everyone to work on the land the Khmer Rouge rapidly increased food production to pre-war levels; in this sense only, they overcame the legacy of the war. But they murdered, starved, or worked to death more than one million people in the process, a staggering human cost.

In this context of large-scale material and human destruction, American private voluntary agencies have aided efforts of the long-suffering people of Vietnam and Kampuchea to heal and rebuild. Since the early 1960s these agencies have received millions of dollars in contributions from thousands of Americans in support of their work on all sides of the conflict in Indochina. Given the immense needs of the Vietnamese and Kampuchean people, agencies such as Oxfam America must continue their humanitarian work in the two countries.

The current State Department embargo on humanitarian aid threatens the ability of the private agencies to meet these needs in Vietnam and Kampuchea as they do in the other parts of the world. Over the past two or three years the staff of the agencies concerned have begun to discuss this challenge to their work. The staff involved in these discussions have agreed on certain basic principles regarding the work of American private voluntary agencies in embargoed countries. These principles may be summarized as follows:

1) *Independence.* To fulfill their humanitarian missions, private agencies must be able to work independent of partisan political and foreign policy concerns of governments. These governments often conflict, which results in human suffering. Humanitarian aid agencies have enough experience to know that suffering occurs on both sides of conflicts. American agencies require independence to meet the needs of the victims according to humanitarian, not political, criteria.

2) *Reconciliation.* Foreign policy defines "enemies" and seeks means to defeat these enemies militarily, economically, and politically. Yet as conflict subsides or alliances shift and interests begin to coincide, the need for reconciliation becomes paramount. By working on a people-to-people basis, even in enemy countries, private agencies lay the foundation for future reconciliation between governments by building trust between people. Preventing agencies from working in embargoed countries aborts this process which,

even if viewed narrowly, is in the long-term interest of the United States.

3) *Aid for Self-Reliance.* All aid, whether provided in an emergency or as part of the on-going struggle for development, should ideally contribute to the long-term self-sufficiency of the beneficiaries. By allowing only emergency aid, and not allowing aid that could provide a way to build independence in case a disaster strikes, the U.S. government helps guarantee future continued emergencies in embargoed countries. To deny aid for self-reliance is to ensure that poor countries like Kampuchea hover on the precarious edge between bare subsistence and serious famine with all its attendant suffering and death. Private humanitarian agencies recognize their responsibility to do more than feed the victims of drought or war; these victims must have the means to recover sufficiently. This inevitably involves aid which works towards long-term solutions.

The U.S. government, specifically the State Department, should recognize these principles and allow the private agencies to do their humanitarian work independent of the foreign policy of the United States. An interpretation of U.S. export-import controls which recognizes these principles will recognize the importance of the work of American international relief and development organizations in lessening human suffering and meeting human need in all developing countries, including those designated enemies of the United States.

Oxfam America and the other aid agencies working in Vietnam and Kampuchea lack the resources to meet even a tiny portion of the humanitarian needs of these countries. Ending the embargo on humanitarian aid, therefore, would have a significant, but limited impact. The primary obstacle to recovery is the continuing political stalemate in the region. As long as Vietnamese troops remain in Kampuchea with diplomatic and military support of the Soviet Union, as long as remnants of the Khmer Rouge, with its murderous leadership intact, and its coalition partners resist the Vietnamese occupation with diplomatic and military support of China, ASEAN, and the United States, there will be no recovery for the people of the region. This stalemate, cynically encouraged by the superpowers in pursuit of narrow strategic objectives, continues to prevent the people of Vietnam and Kampuchea from devoting all their energies to the urgent and daunting task of healing and rebuilding their societies after decades of destruction.

The stalemate has a serious human cost. In poor, bare hospitals in the western part of Kampuchea, children wounded in the border clashes lie in shock. At a textile factory in Phnom Penh, ingenious, determined workers must make by hand metal pieces to repair aging looms that are breaking down past repair one by one, so that now only 200 of 325 still function. Throughout the country the shadow of another serious food shortfall haunts the fields where widows, children, and the few surviving men work long hours, worried at the scantiness of the rice grains they harvest—rice grown without fertilizer, without proper irrigation, without machinery, without enough draft animals—with nothing but the hardest labor. After more than a decade of death and turmoil, the suffering of the Khmer people continues, as a direct consequence of the global machinations of superpowers.

The United States, which has used its power so destructively in Vietnam and Kampuchea, has the opportunity to use its power creatively to end this debilitating political stalemate. The government of the United States, with the American people, must begin working urgently for reconciliation with the governments of Vietnam and Kampuchea. Having attempted repeatedly without success to reach strategic ends through the use of violence and embargoes to punish and to bleed, the United States must take the first step towards building a new relationship with Indochina. After so much violence and suffering there remains a road not taken—the path of negotiation, reconciliation, and eventual respect and friendship. Without this reconciliation, which the United States must dare to initiate, the Vietnamese and Kampuchean people will continue to be victims of isolation, war, and violence. They have been our "enemies" long enough. The time cannot come soon enough when the war has truly ended in Vietnam and Kampuchea.

Glossary

ASEAN

The Association of Southeast Asian Nations: Thailand, Singapore, Malaysia, Indonesia and the Phillipines, recently joined by Brunei.

Democratic Kampuchea (DK)

Official name of the country under Khmer Rouge rule from 1975 to early 1979.

Democratic Republic of Vietnam (DRV)

The official name of the government founded by Ho Chi Minh and other anti-French revolutionaries in 1945. It governed North Vietnam between 1954 and 1976.

Hectare

10,000 square meters, or approximately 2.5 acres.

Heng Samrin

Army officer under the Khmer Rouge. Participated in revolt against Pol Pot and assumed power of the People's Republic of Kampuchea after the Vietnamese invasion in December 1978.

Ho Chi Minh City

The new name for an administrative district which includes the former southern capital, Saigon, plus its neighboring "Chinatown," Cholon, and a surrounding rural area.

Hoa

The Vietnamese term for ethnic Chinese.

Hoc Tap

The official journal of the Vietnamese Workers Party. The party took the name Vietnamese Communist Party after the reunification of the country, and the journal was renamed *Tap Chi Cong San* [Communist Review].

Khmer

The dominant ethnic group of Kampuchea; also the name of the Kampuchean language.

Khmer Rouge

Kampuchean communist movement which fought the civil war

against the Lon Nol government and governed Kampuchea from 1975 to early 1979.

Lon Nol

Head of State of the Khmer Republic from 1970 until overthrown by the Khmer Rouge in 1975.

Metric Ton

1,000 kilograms, or about 2,200 pounds.

Pol Pot

Revolutionary name of Saloth Sar. Prime Minister of Democratic Kampuchea from 1976-1978. Now leader of the Khmer Rouge resistance to the People's Republic of Kampuchea.

Provisional Revolutionary Government of the Republic of South Vietnam (PRG)

The resistance government established by southern revolutionaries in 1969. The core leadership of the PRG was provided by the National Liberation Front (NLF). The core of the NLF, in turn, was the People's Revolutionary Party, southern branch of Vietnam's Workers (Communist) Party. The party, front and PRG were all referred to by many westerners as the "Viet Cong."

Republic of Vietnam (RVN)

The U.S.-backed southern government, based in Saigon, often referred to in the West as "the South Vietnamese."

Sihanouk, Prince Norodom

King of Kampuchea from 1941 to 1955. Abdicated in 1955 and ruled as Prince until 1970. Head of government in exile during armed struggle against Lon Nol. Briefly Head of State under Khmer Rouge but forced to retire in 1976. One of three leaders of resistance coalition against the Heng Samrin government.

Socialist Republic of Vietnam (SRV)

The official name of the government of Vietnam since reunification in 1976.

Chronology of Vietnam

1945: Japan surrenders at the end of the Second World War. A government led by the revolutionary Ho Chi Minh declares the independence of Vietnam and begins to take power from the defeated Japanese.

1946: The French attempt to reassert their colonial control over Vietnam. Ho's resistance government withdraws to the countryside and the "First Indochina War" begins.

1950: The Soviet Union and the People's Republic of China recognize Ho's Democratic Republic of Vietnam. U.S. military aid to the French in Indochina is stepped up after the outbreak of the Korean War.

1954: French forces are defeated at the remote mountain garrison of Dien Bien Phu. The Geneva Conference on Indochina ends the French war and temporarily divides the country at the 17th parallel. The DRV controls the North. Ngo Dinh Diem is premier in the South; Emperor Bao Dai is the southern head of state.

1955: Diem refuses to hold elections for a government of all Vietnam, promised in the 1954 Geneva accords.

1960: The National Liberation Front is formed in the South to lead the resistance against Diem and the United States, which supports his government.

1963: Diem and his brother Ngo Dinh Nhu are assassinated in a military coup which has at least tacit U.S. approval.

1964: Congress approves the Gulf of Tonkin Resolution, later cited as authorization for the U.S. military involvement in Vietnam.

1965: U.S. troop strength in Vietnam reaches nearly 200,000 by the end of the year.

1968: A National Liberation Front offensive at Tet (the lunar New Year) includes attacks on most major population centers. President Lyndon Johnson announces he will not run for another term. In October, he suspends bombing of the North.

1969: Peace talks in Paris include representatives of the United States, the Republic of Vietnam (Saigon), the Democratic Republic of Vietnam (Hanoi), and the Provisional Revolutionary Government of the Republic of South Vietnam (PRG—formed the same year by the National Liberation Front and other resistance supporters).

1973: The Paris Agreement on Ending the War and Restoring Peace in Vietnam is signed, January 27. President Richard Nixon declares that the United States continues to recognize Saigon as the sole legitimate government in the South, contradicting terms of the agreement. U.S. troops leave Vietnam, but advisors remain in an expanded Defense Attache Office.

1975: A major revolutionary offensive topples the Saigon government. PRG and DRV troops enter Saigon April 30.

1976: Vietnam is formally reunified as the Socialist Republic of Vietnam.

1977: Under the administration of President Jimmy Carter, the United States and Vietnam begin discussions of normalizing relations. Vietnam insists that the United States must help "heal the wounds of war" as provided in the Paris Agreement. Vietnam-Kampuchea tensions build, with major skirmishing along the border.

1978: Vietnam joins COMECON. The United States calls off normalization talks, even though Vietnam has dropped its insistence on U.S. reconstruction assistance. Vietnam and the Soviet Union sign a friendship pact. The United States and China establish diplomatic relations. The number of ethnic Chinese leaving Vietnam rises dramatically as tensions between Vietnam and China mount. China halts its aid projects in Vietnam. Vietnamese troops move into Kampuchea to topple the Democratic Kampuchea government.

1979: China attacks along Vietnam's northern border in response to the Vietnamese invasion of Kampuchea. Vietnam's rice production makes a modest recovery after two years of declining production. The Central Committee adopts liberalized guidelines for economic policy.

1980: Vietnam's first full post-war census discloses that the country has a population of 52.7 million, and that it is growing at the rate of about 1 million each year.

1982: A congress of the Communist Party formally ratifies liberal economic policies and a slow-down on collectivizing agriculture. Debates on economic policy continue.

1983: Vietnam announces a record post-war harvest of "nearly" 17 million tons of rice and other staples. Observers say rice production was probably over target, but low production of other staples probably held the total to some 16.7 million tons, slightly below the plan targets.

Chronology of Kampuchea

1953: Kampuchea gains independence from France under the leadership of King Norodom Sihanouk.

1954: The Geneva Conference on Indochina recognizes Kampuchea's neutrality.

1955: Sihanouk abdicates in favor of his father to become the country's principle political leader as Prince.

1963: Sihanouk ends American aid programs. Left-wing opponents of Sihanouk, including Pol Pot, leave Phnom Penh to join the resistance in the Kampuchean countryside.

1965: Sihanouk breaks diplomatic relations with the United States.

1967: Sihanouk's military crushes a peasant revolt in Battambang province. More left-wing opponents flee to the countryside, including the remainder of the future leadership of the Khmer Rouge.

1969: The secret US bombing of Kampuchea begins.

1970: *March 19:* Sihanouk is overthrown in a coup by Prime Minister Lon Nol.

March 23: From Peking Sihanouk announces the formation of a United Front with his former enemies, the Khmer Rouge, to oppose the Lon Nol regime.

April 30: US and South Vietnamese troops invade Kampuchea, without Lon Nol's knowledge or approval, in order to attack bases of the North Vietnamese and National Liberation Front. US troops withdraw on June 30.

1973: *January 27:* Paris Peace Agreement signed. Article 20 calls on all foreign countries to "put an end to all military activities in Cambodia."

February 8:	US bombing of Kampuchea resumes after a halt since January 27.
August 15:	US bombing ceases under orders from Congress.
1975: April 17:	The Khmer Rouge capture Phnom Penh and begin emptying major towns and cities.
September 9:	Sihanouk returns to Phnom Penh as Head of State of Democratic Kampuchea.

1976: Sihanouk resigns as Head of State on April 4. He remains under house arrest for the remainder of the Pol Pot period.

1977: Heavy border skirmishes begin between Kampuchea and Vietnam.

1978: The Vietnamese invade Kampuchea on Christmas day in support of a small United Front for National Salvation.

1979: Vietnamese troops capture Phnom Penh on January 7 and install Heng Samrin as Head of State of the People's Republic of Kampuchea. Fighting continues between the Vietnamese and retreating Khmer Rouge troops. By mid-year famine threatens. Aid effort begins, initially from Vietnam and the Soviet bloc, then from international and Western aid agencies. In October the UN votes to continue seating the Khmer Rouge as the representative of the people of Kampuchea.

1982: To head off mounting dissatisfaction with the continued Khmer Rouge presence in the UN, the Coalition Government of Democratc Kampuchea is formed in exile by former enemies Khieu Samphan (Khmer Rouge), Son Sann (Khmer People's National Liberation Front) and Sihanouk. This loose coalition retains the UN seat in October.

1983: The Heng Samrin regime controls the bulk of the territory and population of Kampuchea. The Coalition controls some border enclaves near Thailand and enjoys the UN seat and recognition by much of the world. This stalemate shows no sign of being broken.

State Department Policy Statement on Licensing

The following memorandum constitutes the only available written statement on U.S. licensing policy related to Vietnam and Kampuchea. It grew out of a policy review conducted by the staff of the Vietnam/Laos/Kampuchea Desk at the State Department in late 1980 and early 1981. It was approved in February, 1981, and released to the voluntary agencies at the time of the denial of a license to the Mennonite Central Committee to ship 250 tons of donated wheat to Vietnam in May of that year. Although nearly three years have elapsed since the policy was adopted, State Department officials told Oxfam America staff that no formal updating of this memorandum had taken place as of December, 1983.

The following is the text of the policy statement:

U.S. Policy on Licensing Shipments to Kampuchea and Vietnam

I. Kampuchea

A. U.S. Funded or Multilateral Projects

The USG has funded a number of projects designed to provide emergency relief to Kampucheans in danger of starvation or disease. Public Law 96-110 which authorizes appropriations for assistance in Kampuchea states that the assistance "shall be for humanitarian purposes and limited to the civilian population, with emphasis on providing shelter, transportation for emergency supplies and personnel, and similar assistance to save lives."

As the situation in Kampuchea improves there will be a declining need for emergency shipments of food, clothing, medicine and related supplies. There will be a corresponding need to examine more closely proposed projects involving U.S. funds to ensure that they satisfy the statutory requirement of "providing emergency relief," and do not constitute projects for the rehabilitation or development of Kampuchea. The humanitarian and political objective of US-funded relief is to ensure the survival of the Khmer race and to provide the means for the Khmer people to live in their country, thereby reducing the pressure of Khmer refugees on neighboring Thailand. While it is accepted that the relief assistance will inadvertently assist the Vietnamese-imposed regime to develop its administration, no assistance should be given beyond the subsistence level which contributes toward consolidation of the Heng Samrin regime's control over Kampuchea.

Rehabilitation projects are defined here as those designed to restore the situation in Kampuchea to that which existed before the Vietnamese invasion in December 1978 (e.g. spare parts for existing machinery, boats and nets to replace those lost or destroyed, supplies to enable medical facilities previously in existence to resume operations). Development projects are defined as those designed to begin new enterprises or operate old ones at previously unattained levels. It is recognized that there are not always clear-cut distinctions between relief, rehabilitation, and development. Circumstances can alter, so that projects to supply seed, agricultural implements and nets could be justified during a severe food shortage, but would be considered to be rehabilitation projects once a level of subsistence is close, and additional food production would lead to surpluses. In circumstances where rehabilitation projects contribute toward the goals of relief, as defined above, the use of U.S. funds can be authorized on a case-by-case basis. U.S. funds cannot be authorized for development projects in Kampuchea.

The distinction between relief and rehabilitation can be illustrated in the case of transportation equipment. In the early stages of the Kampuchea relief efforts, lack of transportation, including airport and seaport facilities, rail lines and trucks, severely impeded relief efforts. International aid was given in all of these areas to the point where relief supplies finally could be transported to the people in need. As Kampuchean agriculture improves, the need for transportation to move goods may increase, but this is not a relief need and should not be funded by the USG.

B. Private and International Organization Donors

Licenses for goods donated by private and international aid organizations for relief purposes in Kampuchea generally will be granted when the items are for humanitarian purposes, especially items which satisfy basic human needs. At the present time, the criteria for approval of export licenses will be essentially the same as those used for US-funded relief. Such categories as food, medicine, medical supplies, basic agricultural supplies, equipment necessary for transport of relief supplies, and clothing will usually be approved so long as they are needed for relief work. We will consider other items if it can be shown that they are necessary for relief operations.

Donations made for rehabilitation and development projects will generally not be approved except where the foreign policy or other interests of the USG are served. Items or projects which tend to strengthen the Vietnamese-supported regime in Kampuchea will not be approved.

II. Vietnam

General Policy:

Any assistance to Vietnam, beyond the most elementary humanitarian aid, cannot be approved at this time because it would directly or indirectly assist the SRV in pursuing its adventure in Kampuchea.

The severe economic hardship of the Vietnamese people is recognized, including hardship brought on by natural disaster, but it also is recognized that the Vietnamese government has within its power the ability to alleviate this hardship by ending its diversion of resources from economic development to military conquest.

The private groups which provide aid to Vietnam serve the U.S. national interest in maintaining a channel of communication to the officials and people of Vietnam. The existence of this private aid program also provides a means by which the USG can, when the time arises, send a positive signal to Vietnam by permitting an increase in the level of private assistance. However, until the time comes to send that positive signal, the private aid should be maintained at a token level.

A. USG Funded Projects

Legislation prohibits expenditures of U.S. funds for economic assistance to Vietnam. Relief can be authorized under PL 480. In the event of a documented evidence of large-scale natural disaster in Vietnam, the provision of a token amount of disaster relief can be considered in the light of the general diplomatic situation at that time, particularly vis-a-vis Kampuchea.

B. Private and International Organization Donors

Licenses will generally not be granted except where humanitarian factors are involved or where foreign policy or other USG interests are served. Applications for validated licenses will generally be denied except that non-commercial exports to meet emergency needs will be considered on a case-by-case basis.

Oxfam America
Licensing Experience

A. *Vietnam*

1. *Tich Giang Seed Cooperative, $28,000.* Seed processing and storage equipment for a small seed cooperative. Treasury Department license application filed March 10, 1982. License denied July 30, 1982. Appeal submitted August 11, 1982. Appeal rejected October 5, 1982.
   PROJECT STATUS: Never implemented.

2. *Pho Hien Bee Culture Cooperative, $18,000.* Bees wax and wire screening for beekeeping cooperative which supplies honey as a food supplement to hospitals and day care centers. Treasury Department license application filed March 10, 1982. License denied July 30, 1982. Appeal submitted August 11, 1982. Appeal rejected October 5, 1982. After discussion with State Department, Treasury Department license application re-submitted January, 13, 1983. License granted May 5, 1983.
   PROJECT STATUS: Implementation under way.

3. *Tu Loc District Health Services, $24,000.* Medicines, laboratory equipment, and medicine processing equipment to increase traditional medicine production. Treasury Department license application filed March 10, 1982. License granted July 30, 1982.
   PROJECT STATUS: Implementation delayed when Vietnamese balk at implementing only one of four projects. Implementation under way.

4. *Cantho University Agricultural Program, $60,000.* Audio-visual equipment and materials for training program; seed storage equipment for a seed multiplication program. Treasury Depart-

ment license application filed March 10, 1982. License denied July 30, 1982. Appeal submitted August 11, 1982. Appeal rejected October 5, 1982. PROJECT STATUS: Never implemented.

5. *Typhoon Nancy Relief, $60,000.* Emergency food relief to victims of Typhoon Nancy. Treasury Department license application filed November 5, 1982. License granted December 20, 1982. PROJECT STATUS: Implementation completed.

B. *Kampuchea*

1. *Battambang Phosphate Factory, $40,000.* Spare parts for phosphate rock crusher at the factory. Commerce Department license application filed December 30, 1981. License granted February 28, 1982. PROJECT STATUS: Implementation completed.

2. *General License, $1 million.* Basic agricultural supplies to support agricultural recovery. Treasury Department license application filed January 4, 1982. Application withdrawn June 2, 1982 after request for itemized list of supplies and destinations. Decision made to submit applications on project-by-project basis.

3. *Prey Veng Irrigation Stations, $19,000.* Cement and steel reinforcing rods to repair canals and gates. Treasury Department license application filed March 25, 1982. License granted July 9, 1982. PROJECT STATUS: Implementation completed.

4. *Battambang Phosphate Factory, $58,000.* Transportation equipment. Treasury Department license application filed March 25, 1982. License granted July 9, 1982. PROJECT STATUS: Implementation completed.

5. *Battambang Rice Mills, $65,000.* Spare parts to repair 10 large rice mills. Treasury Department license application filed November 15, 1982. License granted January 7, 1983. PROJECT STATUS: Implementation completed.

6. *Battambang Phosphate Factory, $16,000.* Plastic bags for storage and distribution of phosphate fertilizer. Treasury Department

license application filed April 1, 1983. License granted April 20, 1983.

PROJECT STATUS: Implementation completed.

7. *Prey Veng Traditional Irrigation Pump Repair, $38,300.* 300 tool kits for the repair of traditional wooden irrigation pumps. Treasury Department license application filed April 15, 1983. License granted June 20, 1983.

PROJECT STATUS: Implementation under way.

8. *Banan Agricultural Station, $9,650.* Spare parts for power tillers and irrigation pumps, and two hand-pulled rice seeders for a small agricultural station. Treasury Department license filed April 15, 1983. License granted June 20, 1983.

PROJECT STATUS: Implementation under way.

9. *Battambang Phosphate Factory, $46,000.* Transport equipment to enable more rock to be transported efficiently to the factory. Treasury Department license application filed April 15, 1983. License denied June 20, 1983. Appeal submitted September 9, 1983. License granted January 3, 1984.

PROJECT STATUS: Implementation under way.

10. *Solar Irrigation Pumps, $30,062.* Ten small-scale solar irrigation pumps for rice irrigation. Commerce Department license application filed April 21, 1983. License denied October 24, 1983. Appeal submitted November 8, 1983. Appeal still pending.

PROJECT STATUS: Not implemented pending decision on appeal.

11. *Battambang Rice Mills, $35,000.* Spare parts and other supplies to repair a steam-powered rice mill, fueled by rice husks. Treasury Department license application filed August 31, 1983. License denied October 20, 1983. Appeal submitted November 14, 1983. Appeal rejected January 9, 1984.

PROJECT STATUS: Never implemented.

C. *Summary*

1. *Vietnam*
   Applications filed: 5, all with Treasury Department
   Applications DENIED: 3 valued at $106,000

Applications initially GRANTED: 2 valued at $84,000

GRANTED after appeal: 1 valued at $18,000

Total $ value licensed projects: $102,000

Total $ value denied projects: $88,000

Average time from application to final decision: 6 months, 3.5 weeks.

2. *Kampuchea*

Applications filed: 11, 9 with Treasury Department, 2 with Commerce Department

Applications DENIED: 3 valued at $111,062

Applications initially GRANTED: 7 valued at $245,950

Applications WITHDRAWN: 1 valued at $1 million

Applications GRANTED AFTER APPEAL: 1 valued at $46,000; 1 appeal still pending

Average time from application to final decision: 3 months, 3 weeks.

NOTE: All Kampuchea denials occurred after June 1983, indicating increasingly restrictive State Department policy.

The U.S. Aid Pledge

The following is a section of a report submitted by Senator George McGovern to the Senate Foreign Relations Committee in March 1976. Senator McGovern wrote the report after a visit to Vietnam and in it he summarizes his views and the views of the Vietnamese government on the state of United States-Vietnam relations at that time. The section reproduced here covers the U.S. aid pledge made and then withdrawn by President Nixon during the negotiations from October 1972 to July 1973.

THE PROCESS OF AGREEING TO AID

In the course of a general description of events immediately preceding and immediately following the Paris Agreements, Xuan Thuy [Chief North Vietnamese delegate to the Paris Peace talks] revealed that there was also an agreement on a specific level of American aid.

The October, 1972, draft agreement, he said, was cabled by Secretary Kissinger to President Nixon. According to Thuy, Mr. Nixon "answered that he would agree, and he made an appointment in October that it would be signed."

> Later he asked for the signing to be postponed for a later time, and he demanded that some agreed upon provisions be changed. We told them that they could modify details but they could not, it was impossible, to modify the essentials.

In an earlier dinner conversation Xuan Thuy said Secretary Kissinger had made a "definite commitment" to sign the agreement by the end of October, but that he "swallowed his promise." Thuy said the Secretary had been asked directly if he could speak for Saigon in making that commitment, and he quoted the response as "I would not be here if I couldn't."

When President Nixon requested a postponement in the signing, Xuan Thuy said they received word from "confidential sources" that the United States was, in fact, backing out of its commitment to the agreement. That, he said, is why the DRV made its public announcement that an agreement had been reached and that it was to be signed by the end of October—so the public "would know an agreement was made, and that it was not we who were blocking it."

When the negotiations broke up it was the North Vietnamese understanding, according to Xuan Thuy, that the teams were reporting to their respective governments. Thuy was still in Paris meeting with Ambassador Bruce when the December, 1972, bombing began. Le Duc Tho had arrived in Hanoi barely two hours before.

During the Christmas bombing, Xuan Thuy said the White House sent word that they wanted to meet again. But he recalled that "our government said that we would only meet again under the condition that the United States must stop the bombing." The bombing stopped and the agreement was signed on January 27.

On the question of aid, Thuy said the first agreement included a commitment on the part of the United States to "participate in the healing of the war wounds and the reconstruction of Vietnam." Then, at the time of the January agreement, Thuy said Mr. Nixon sent a memorandum.

In his letter to Premier Pham Van Dong, Nixon said the United States would participate in the healing of the war wounds to postwar Vietnam and would give $3.25 billion in economic aid. He also proposed the establishment of a special economic commission. We agreed.

However, after discussions in Paris, there was no result. We concluded that the American side just promised reparations but in fact they didn't want to implement the promise.

Information on this specific aid agreement had also been supplied by Deputy DRV Foreign Minister Phan Hien to the members of the House Select Committee on Missing Persons in Southeast Asia during their December visit to Hanoi. The Nixon message, dated February 1, 1973, was described as stating that the United States would contribute to the reconstruction of North Vietnam "without any political considerations," that the U.S. contribution would be $3.25 billion over a five year period, with other forms of aid to be agreed upon by the two sides, that details were to be reviewed by the two governments, and that a Joint Economic Commission would be formed to complete negotiations on the details of an aid agreement.

Pham Van Dong responded immediately with a message confirming all the points in the Nixon message.

The Joint Economic Commission described in this exchange began meeting in Paris on March 15, 1973, and, according to information supplied to the House Select Committee, it did prepare a draft agreement:

> . . . the Commission had actually reached agreement on the total amount of grant aid to be provided . . . the percentage to be spent in the United States (85 percent) and in third countries (15 percent), the list of commodities to be purchased over the entire five years, and the commodities to be purchased during the first year.
>
> . . . the United States was expected to play a central role in the reconstruction of North Vietnam, with the emphasis on industrial plants and commodities, infrastructure, and energy. The five year plan provides for plants for prefabricated houses, plumbing fixtures, sanitary porcelain ware, cement, sheet glass, chipboard, synthetic paint, and a steel mill with an annual output of one million tons. The contribution to energy development included a thermal power station with a capacity of 1,200 megawatts, a high tension electrical equipment plant with an annual output of 3,000 tons of high tension copper cable. In addition, the agreement included a provision of a vast array of equipment for port reconstruction and water, road, and rail transport, and for agriculture.

The Commission, including three delegates from each side, met until President Nixon suspended U.S. implementation of the Paris Agreement in April. It met again in June and July. But on July 23, when the detailed aid agreement was scheduled to be signed, the United States instead broke off all talks indefinitely.

The existence of the letter from President Nixon to Pham Van Dong—and its existence has been confirmed by State Department spokesmen in recent weeks—creates serious circumstantial doubts about the Administration's assertions at the time that they had agreed to no specific aid program in the context of the Paris Agreement. Believing that in retrospect requires acceptance of one of two highly unlikely events: Either that President Nixon set the $3.25 billion figure on his own and voluntarily forwarded the letter, or else that somehow the two sides worked feverishly between January 27 and February 1 to agree upon the specific amount and the terms that were included in the Nixon message. More likely the Nixon memorandum itself was the product of earlier hard bargaining and an undisclosed understanding reached before the Paris Agreement was signed. That, too, has been confirmed privately by knowledgeable sources in the Administration.

In turn, the Nixon memorandum and the negotiating context described by Xuan Thuy both undercut the Administration's claim—which was wobbly enough at the time—that the Christmas bombing produced major negotiating results for the United States. As Xuan Thuy described it, the bombing halt was not a magnanimous gesture on the part of the United States, but a North Vietnamese precondition to resuming the discussions. The bombing could not have long continued anyway, because, at the same loss rates, the entire fleet of B-52 bombers assigned to Southeast Asia would have been lost in about 90 days time. Then, as a consequence of the added damage inflicted upon North Vietnam, the Administration had to agree to a reconstruction aid figure much higher than anyone had supposed (no specific aid figure was ever formally requested of the Congress, but the sum discussed in widespread news accounts was $2.5 billion for North Vietnam).

Further, the details of the aid discussions carried out pursuant to the Nixon letter, and the existence of an actual draft aid agreement, shed new light on the question of who was responsible for the eventual collapse of the ceasefire. Up until July 23, 1973, the Vietnamese had every reason to believe that the Administration planned to provide reconstruction aid. That was another strong incentive for them to maintain a purely defensive military posture. But on July 23, the Nixon Administration broke off the aid talks, and seemed to be adding a new condition—a requirement that the Vietnamese somehow arrange a ceasefire in Cambodia—to the terms of the Paris Agreement. This was also a deviation from the Nixon letter, which promised the aid "without any political conditions." By the most knowledgeable accounts, it was only then—when the Nixon Administration, as well as the Thieu government, had demonstrated bad faith—that the Vietnamese Communists began to prepare for a more aggressive military response to ARVN incursions.

I did not receive the impression in Hanoi that the disclosures on the Nixon letter and the draft aid agreement were made to reflect the current North Vietnamese position on the amount of aid they would expect from the United States if the United States were to accept its obligations under Article 21. That provision of the Paris Agreement came up repeatedly in my meetings with both DRV and PRG officials, yet the Nixon message was not mentioned by anyone other than Xuan Thuy. Moreover, it was then raised not in connection with our discussions of Article 21, but in the context of a description of the negotiating process in late 1972 and 1973.

To be sure, the extent of the destruction to North Vietnam was brought home forcefully. I was told that up until July, 1972, the material loss amounted to more than $6 billion, exclusive of the

147

Christmas bombing. We saw the Bach Mai hospital which we were told had been bombed three times—June 27, December 19, and December 22—in 1972. On the latter occasions, we were told, more than 100 bombs struck, and 28 medical personnel were killed. The hospital has been rebuilt since the war, but slides were used to demonstrate its earlier condition. Our guides told us aid for rebuilding the hospital had come from China, and that $1 million in private American contributions had been sent through Medical Aid for Indochina. We also saw sections of dike four kilometers from Hanoi which had been bombed and rebuilt. The earliest atttacks on the dikes were in August of 1966, we were told, and then again in 1967, in 1968, and during the Christmas bombing in 1972.

The dikes are obviously crucial not only in North Vietnamese agriculture but to prevent flooding of populated areas. We crossed the two-kilometer bridge which handles motor, rail, bicycle, and foot traffic across the Red River separating Hanoi from Gia Lam airport. It had been destroyed three times during the war. We stopped in a residential area, the Kham Thien District near the center of Hanoi, which was described as the site of the most severe human losses in the Christmas bombing—270 people killed on the night of December 26, 1972. We were told that thousands of unexploded bombs still exist in the rural areas of North Vietnam, and that they still cause occasional fatalities. Premier Pham Van Dong described massive damage to factories, communications and transportation networks, schools, housing, and hospitals. Xuan Thuy said that if we had been there in early 1973, we would have seen that all railroads and roads had been damaged and that all major bridges had been destroyed. He reported that much has been restored, but not the railroads. Now school construction permits study in two shifts a day now, instead of three.

Notwithstanding these descriptions, I came away with the impression that the North Vietnamese remain flexible on the size and nature of any potential American aid program. Premier Pham Van Dong put it in terms of the American people having "some part" in rebuilding the country. He did not mention the Nixon memorandum or the 1973 proceedings of the Joint Economic Commission; instead he said that "The exact sum is not mentioned in the Paris Agreement, but it is a matter of honor, responsibility and conscience."

APPENDIX VII
For Further Reading

A. *US Export-Import Controls*

Bergeron, Pierre. *Export-Import Controls: Their Nature and Accountability, Approaches to Fostering U.S. Recognition of Foreign Governments.* New York: Church World Service, 1980. Available from Church World Service, 475 Riverside Drive, New York, N.Y. 10115.

B. *Vietnam*

Emerson, Gloria. *Winners & Losers: Battles, Retreats, Gains, Losses and Ruins from a Long War.* New York: Random House, 1976.

Fall, Bernard B. *Viet-Nam Witness: 1953-66.* New York: Praeger, 1966.

Fall, *Last Reflections on a War.* Garden City, NY: Doubleday, 1967.

Isaacs, Arnold. *Without Honor: Defeat in Vietnam and Cambodia.* Baltimore: Johns Hopkins University Press, 1983.

Luce, Don and John Sommer. *Viet Nam: The Unheard Voices.* Ithaca: Cornell, 1969.

Maclean, Michael. *The Ten Thousand Day War.* New York: Avon, 1981.

The Pentagon Papers. New York: *The New York Times*, 1971.

C. *Kampuchea*

Chandler, David P. *A History of Cambodia.* Boulder, Colorado: Westview Press, 1983.

Delvert, Jean. *Le Paysan Cambodgien.* Paris and The Hague: Mouton & Co., 1961.

Kampuchean Inquiry Commission. *Kampuchea in the Seventies: Report of a Finnish Inquiry Commission.* Helsinki: Kampuchean Inquiry Commission, 1982.

149

Kiernan, Ben and Chantou Boua, eds. *Peasants and Politics in Kampuchea, 1942-1981*. London: Zed Press, 1982.

Ponchaud, Francois. *Cambodia Year Zero*, translated by Nancy Amphoux. Hammondsworth, England: Penguin Books Ltd., 1978.

Shawcross, William. *Sideshow: Kissinger, Nixon, and the Destruction of Cambodia*. New York: Simon and Schuster, 1979.

Vickery, Michael. *Cambodia 1975-1982*. Boston: South End Press, 1984.

D. *Periodicals on Regional Events and Issues*

Far Eastern Economic Review, published weekly in Hong Kong. Hong Kong address: GPO Box 160, Hong Kong. US Mailing Agent: Datamovers, Inc., 38 W. 36th Street, New York, N.Y. 10018.

Indochina Issues, published monthly by the Indochina Project, Center for International Policy, 120 Maryland Avenue, N.E., Washington, D.C., 20002.

Southeast Asia Chronicle, published by the Southeast Asia Resource Center, P.O. Box 4000-D, Berkeley, California 94704.